The Automated Office
An Introduction to the Technology

The Automated Office
An Introduction to the Technology

By William Saffady

National Micrographics Association
8719 Colesville Road
Silver Spring, Maryland 20910
301/587-8202

RS17-1981

International Standard Book Number 0-89258-072-0

© 1981 by the **National Micrographics Association**
                8719 Colesville Road
                Silver Spring, Maryland 20910
                301/587-8202

All rights reserved
Printed in the United States

Edited by Gene A. Plunka
Book design by Jean W. Farmer
Cover and dust jacket design by Bob Petz Design, Silver Spring, Maryland
Composition by Monotype Composition, Baltimore, Maryland
Printing and binding by BookCrafters, Inc., Chelsea, Michigan

Contents

Foreword   vii

1 **Office Automation: An Introduction**   1
 An historical review of automation / Office automation today / Automated office technology

2 **Word Processing**   21
 Dictation systems / Automated text editing / Text editing and related technologies

3 **Computers and Office Automation**   51
 Online terminals / Computers and management information / Small business computers

4 **Micrographics**   89
 Source document microfilming / Computer output microfilm / Computer-assisted retrieval (CAR)

5 **Reprographics**   127
 Office copiers and duplicators / Phototypesetting

6 **Electronic Mail and Message Systems**   153
 Electronic mail systems / Image-oriented message transmission / Character-oriented message transmission

7 Integrating and Implementing Automated Office Systems   193
Integrated information systems / Some nontechnological considerations

Glossary   207

Notes   215

Index   235

Foreword

Office automation—the application of technology to the creation, storage, manipulation, retrieval, reproduction, and dissemination of information in an office environment—is currently the most dynamic and fastest growing field of information processing. The evaluation, selection, and implementation of automated office systems is, however, a complex undertaking requiring familiarity with a wide range of technologies, each with its own terminology, underlying concepts, and application characteristics. While there are a growing number of books and articles that describe one or more specific facets of office automation, there has been no single source which provides a comprehensive overview of all major areas of office information processing. This book is intended to fill that void. It is designed for information managers, information systems analysts, data processing professionals, micrographics systems specialists, administrative support specialists, records managers, and others who want an introduction to the fundamentals of automated office technology. The book emphasizes the terminology and concepts that are essential to an understanding of the application of technology to office operations.

The organization of the book is as follows. Chapter One introduces the concept of office automation, emphasizing the motives for the currently intense interest in the automation of office operations. Each of the next five chapters deals with a particular facet of office automation. Chapter Two discusses word processing, including both dictation and text-editing systems. Chapter Three deals with office-oriented computer applications. It begins with a discussion of online terminals and time-shared access to data bases and data banks, and concludes with a description of small business computer systems designed for installation and operation in the office. Chapter Four discusses the role of micrographics in office automation. It emphasizes new developments in micrographics technology which are significant for the management of active information. Chapter Five deals with the often neglected subject of reprographics, including copiers, duplicators, and photo-typesetting systems. A section of that chapter deals with the "intelligent copier," a product that reflects the impact of word processing and computer technologies on a supposedly "mature" technology. Chapter Six discusses electronic mail and message systems—a facet of office automation that is expected to increase considerably in popularity. The chapter begins with an overview of recent developments in telecommunications technology, followed by descriptions of available forms of electronic mail—facsimile systems, TWX and telex, commnicating word processing, communicating optical character recognition, and computer-based message systems. The concluding section deals with the potential of video technology in office operations. Chapter Seven provides a brief discussion of the integration of technologies as well as some nontechnological considerations which bear on the implementation of automated office systems. A glossary, which excerpts definitions used throughout the book, is provided, and the footnotes associated with individual chapters contain suggestions for additional reading.

A number of persons were kind enough to read this book in draft form and make useful suggestions for its improvement. I would like to thank Ellen Meyer, Gene Plunka, Jean Farmer, and Bill Neale of the NMA staff, as well as the Publications Committee of NMA, for their assistance during various stages of the work.

William Saffady
Brooklyn, New York

The Automated Office
An Introduction to the Technology

1
Office Automation: An Introduction

An historical review of automation • Office automation today • Automated office technology

It is almost impossible today to attend a meeting of administrative managers or information systems specialists, or to read their professional or trade publications, without encountering an indictment of the office as an expensive, inefficient workplace. Despite significant increases in the amount and complexity of work to be performed, we are told that offices have changed very little since the mid-nineteenth century. At a time when the need to communicate is expanding, the cost to create and send a typical business letter, which now averages $6.07, has doubled since 1970.[1] Although the office workforce is growing rapidly and its wages are rising at seven to eight percent annually, a recent study indicates that clerical employees spend as much as 18 percent of their time waiting for work.[2] In the expectation of a need to house greater numbers of employees and records, acceptable office space has become increasingly scarce and expensive.[3] At a time when the largest percentage of employees are characterized as "information workers," the office's prevalent information processing methodologies are criticized as time-consuming and ineffective.[4]

In an increasing number of organizations, these problems are considered serious enough to have attracted the attention of top management, and various possible solutions are being explored, one of which—office automation—is the subject of this book. Office automation involves the application of technology to the creation, storage, manipulation, retrieval, reproduction, and dissemination of information in an office environment. For purposes of this book, the office environment is broadly defined as the workplace of white collar workers, including managers and administrators, professional and technical personnel, salespersons, and clerical workers. As such, it is a rare business or other organization that does not have an office component.

1

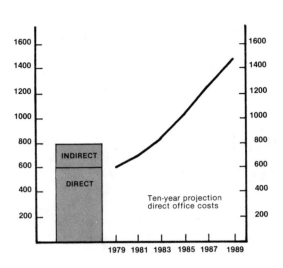

Figure 1-1. In 1979, U.S. businesses spent an estimated $800 billion on office operations, of which $600 billion represented the direct cost of compensation and fringe benefits. The remaining $200 billion of indirect costs covered internal support, space and purchased resources. If current trends are not reversed, the cost of office operations could rise to $1.5 trillion by 1990. (Reprinted from the *Journal of Micrographics,* July/August 1980. Courtesy: Booz, Allen and Hamilton, Inc.)

In its simplest form, office automation aims to increase the productivity of office employees in the performance of existing tasks, thereby reducing the cost of office operations. In its more complex variants, automation permits fundamental changes in the office as a work-place and in the office's role within an organization—changes which some observers contend will lead to the "office of the future." This chapter provides an overview of the automated approach to the problems of the office. It begins with a review of historical patterns of office automation and continues with an analysis of the currently intense interest in the automated office and a discussion of the ways in which the modern office differs from its predecessors. The chapter concludes with descriptions of the technologies and product groups which are the building blocks of office automation. A detailed discussion of those technologies is provided in subsequent chapters.

OFFICE AUTOMATION THEN AND NOW

An Historical Review

While the concept of an "office of the future" dates from the mid-1970s, the current concern about office costs and productivity is hardly new. For over 40 years, industrial engineers and work methods analysts have emphasized the potential for significant improvements both in efficiency and cost-effectiveness to be gained through the scientific study of office procedures.[5] While these specialists typically emphasized the streamlining of paperwork flow and the standardization of clerical work methods through the application of time and motion studies and other industrial engineering techniques, the automation of office activities was not overlooked. Although, as noted in the opening section of this chapter, the office has commonly been viewed as a monolithic entity, a significant amount of mechanization was applied to office tasks prior to the mid-1970s.

From 1900, when use of the typewriter became popular, through the early 1960s, a number of significant pieces of business equipment were assimilated successfully into the office environment, the most intense period of mechanization beginning during World War I. In 1919, the National Association of Office Managers was formed to apply scientific management techniques to office problems. By 1930, at least 30 percent of office clerical employees were using a machine other than a typewriter for at least part of their work. A 1937 report, published in *The International Labor Review*, listed over 30 such devices, including dictation equipment, tabulating machines, sorters, adding machines, and document conveyors. Microfilm cameras and viewing equipment, introduced in the 1920s, were in widespread business use in the 1950s. The photostat machine was first used by insurance and real estate offices in 1910, and by 1960, over 20 different document reproduction processes, including xerography, were available for office use. Teletype machines, introduced prior to World War I, were in widespread use by the mid-1950s. Writing in 1953, sociologist C. Wright Mills said that the pace of this "industrial revolution" was transforming the office much faster than it did factories and noted that while "still in its late infancy, it is clearly a model of the future."[6]

Mills further noted that the introduction of office equipment was accompanied by the standardization and systematization of work steps. Office layouts were redesigned to optimize the flow of work between clerical stations. The division of labor and assembly line methods borrowed from factories were used to increase paperwork production. This systematic approach to the performance of office tasks facilitated further automation when electrical accounting machines and electronic computers were introduced to businesses in the 1960s. From the mid-1960s through the mid-1970s, computers were pervasively applied to the highly routinized clerical work encountered in the "back offices" of accounting and transaction-processing departments. Studies of office automation written during that period are virtually indistinguishable from data processing textbooks.[7] Unlike the business machines of the earlier period with their emphasis on mechanized document handling, this newer equipment emphasized the manipulation of the data content of documents.

There were, however, important enhancements of older technologies during this later period, some of which were to play an important role in emerging automated systems. The IBM Selectric® Typewriter, for example, was introduced in the early 1960s, followed by the IBM Magnetic Tape/

Selectric Typewriter (MT/ST) in 1964, and the first magnetic card typewriter in 1969. Xerographic copiers improved tremendously during the 1960s and early 1970s. Users began experimenting with computer-assisted microfilm retrieval during the late 1960s. Minicomputers became available in the late 1950s, although they were not widely used in business applications until the early 1970s. In a non-technological development which reflected increased acceptance of the inevitability of change, open-plan offices with their readily moveable workstations began to replace conventional office construction in the 1960s and 1970s.[8]

Figure 1-2. Open-plan and landscape interior designs reflect the flexibility and receptivity to change that have characterized office operations since the early 1970s. By using moveable partitions with attached furnishings, interior designers are able to counteract the traditional view of the office as a group of fixed, isolated work spaces. The newest open-plan furnishings are specifically designed for automated offices. (Courtesy: Steelcase Inc.)

Office Automation Today

As the preceding section suggests, the current interest in office automation and the proliferation of automated office products can be viewed as a logical continuation of an historical trend that dates from the early twentieth century. There are, however, important differences in conceptual framework, socio-economic context, and underlying technology which distinguish the most recent developments in office automation from those of their predecessors. Those differences are discussed in this and the following sections.

With regard to conceptual framework, office automation has had one goal since its inception—the improved cost-effectiveness of office operations. Since the mid-1970s, however, manufacturers, vendors, systems analysts, and others concerned with office automation have emphasized new approaches to that goal:

1. *Emphasis on "Unstructured" Activities.* As already discussed, early office applications of computers and other business equipment concentrated on the highly systematized clerical operations characteristic of accounting and customer transaction processing. Information systems analysts describe such "back office" operations as "structured" in the sense that their characteristics are well understood, they occur on a regular schedule, and computerization or other forms of automation are applied to a processing system already in place.[9] By way of contrast, newly developed automated office products are increasingly designed for "front office" applications where managerial and secretarial operations are rarely systematized but are instead a result of external events. Several studies confirm the essentially reactive nature of much office work. Managers, for example, seldom initiate work. They spend most of their time responding to action requests. Secretaries' work patterns generally mirror those of the principals they support. Far from being standardized, secretarial work is varied, and much of it is performed in a highly discretionary manner.

2. *"Paperless" Information Systems.* Through the mid-1970s, office automation relied heavily on successful models established in manufacturing.[10] Paperwork was assumed to be the office "product." In the manner of factory-based systems, office automation emphasized the rapid, more efficient production or other processing of invoices, checks, purchase orders, and similar paper documents. Even office-oriented computer systems which, as noted earlier, emphasized the manipulation of data, did so with the intention of eventually generating reports or other documents. Despite the notoriety accorded the "paperless information systems," paper remains important in the office today, and some automated office product groups—notably, word processing equipment and intelligent copiers—are, in part, designed to facilitate document production. But systems analysts, records managers, and other information management professionals increasingly emphasize the high cost of creating, storing, retrieving, reproducing, and disseminating paper documents. As a result, the new automated office technologies rely on alternative information carriers—notably, magnetic-coated and photosensitive recording materials.

3. *Improved Managerial Effectiveness.* Although clerical employees represent a relatively small and low-paid segment of the white collar workforce, office automation traditionally has emphasized improvements in clerical operations. While the reduction of clerical costs remains an important objective,[11] designers of office information systems increasingly are concerned with the potentially beneficial impact of automation on professional and

managerial costs. Contrary to the academic view of managers as highly efficient, scientific planners, several studies indicate that most managers have little time for the planning they are presumably trained and hired to do.[12] Instead, they are subject to continual interruptions that force them to react to specific crises. The largest percentage of their workday is spent in meetings or on the telephone. While they depend on information for effective decision-making, their information-seeking behavior is unstructured. Even where formal management information systems have been developed, managers continue to rely heavily on personal contacts and other informal information sources. The same information-seeking behavior is characteristic of the professional/technical segment of the white collar workforce.[13] While not rejecting the traditional emphasis on the reduction of clerical costs, automated office systems are increasingly designed to improve the effectiveness and to reduce the cost of managerial and professional personnel.

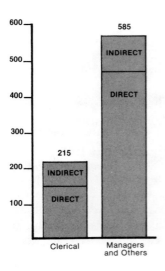

Figure 1-3. While previous efforts at office automation focused on clerical operations, information systems designers are increasingly concerned with the potentially beneficial impact of automation on professional and managerial costs. Clerical costs are estimated to account for only about 27 percent of total office costs. The remainder represents the cost of the so-called "knowledge workers." (Reprinted from the *Journal of Micrographics,* July/August 1980. Source: Booz, Allen and Hamilton, Inc.)

Apart from the calculators used by accountants and technical workers, the non-clerical segment of the office workforce has never been routinely provided with automated "tools." Some of the equipment described in later chapters of this book, however, specifically is intended for managerial rather than clerical use.

4. *The Integration of Technologies.* Earlier phases of office automation relied on methodological approaches and equipment that improved the performance of isolated tasks. Typewriters, copiers, and document conveyors are "stand-alone" devices not designed traditionally to interface with one another or with other types of automated equipment. Similarly, a typical computerized office application in the 1960s or early 1970s operated on data

files designed specifically for that application. Little or no attempt was made to consolidate redundant data in files processed by several application programs or to have the output of one application program serve as input to another.

Today, however, there is an emerging interest in the integration of information processing applications and equipment. This interest is reflected, for example, in the pervasiveness of data base management systems that integrate an organization's information resources. It is likewise expressed in the considerable recent attention given to the integration of technologies—that is, the combined use of several different technologies and types of equipment to meet the needs of a given application. This integration of technologies is currently being accomplished through: (1) the development of multi-function equipment possessing the capabilities of several technologies; (2) the establishment of hardware and software interfaces between "stand-alone" devices; and (3) the use of computers as intermediaries between otherwise incompatible technologies. These approaches recognize the automated office as an information processing system rather than an accumulation of isolated activities to be mechanized on a piecemeal basis.

5. *Changes in Office Structure.* With the exception of the increased routinization of certain clerical jobs, earlier phases of automation had little, if any, impact on the organizational structure of the office. In fact, the major systematization and centralization of office functions preceded automation. Before the proliferation of business machines in the 1920s and 1930s, serious consideration was given to decentralization as a remedy for inefficiency. Automation, however, made the existing centralized office structure workable, thus preserving the status quo.[14] In marked contrast, many of the automated office systems discussed in this book derive their cost-effectiveness from modifications of existing office structures and work patterns. With the introduction of word processing, for example, secretaries often find their jobs radically altered or even enriched. In addition, the existence of computerized data bases accessed from remote terminals threatens the existence of conventional files—long considered a focal point of the traditional office. Similarly, the availability of computer-based electronic message systems may have a significant impact on hierarchical lines of management communication. One of the cost justifications for such systems involves the modification of prevailing patterns of supervision. At the extreme, some information management specialists forecast an eventual dissolution of the office itself as sophisticated communication facilities permit the geographic scattering of employees who will work at multi-function terminals in their homes or other locations.[15]

6. *The New Socio-Economic Context.* Finally, it is important to note that the socio-economic context of office automation is currently much different from that of the 1950s, 1960s, and early 1970s. Historically, the productivity of office employees has attracted little attention. The Bureau of Labor

Statistics, for example, does not distinguish between white collar and blue collar employees in its productivity computations, and it omits the largely office-oriented government workforce from its statistics entirely. Business has typically treated office expenses as undifferentiated overhead, while academic economists traditionally have dismissed white collar workers as unproductive.[16] The office workforce is, however, much larger today and concerns about its productivity can no longer be dismissed. When viewed in the broader context of national economic growth, the need for office-oriented productivity improvement programs appears all the more urgent. These factors are discussed in greater detail in the following section.

Offices and the Productivity Problem

In the United States, productivity, or rather the lack of it, came to the forefront as a national problem in the late 1970s. As an economic concept, productivity is broadly defined as the relationship between the output of goods and services and the input of labor, capital, and natural resources. The U.S. Bureau of Labor Statistics regularly reports several measures of national productivity.[17] The simplest, *labor productivity*, is calculated by dividing the Gross National Product (GNP)—the market value of all goods and services produced by the nation's economy—by the number of persons or employee hours in the workforce. *Total productivity*, a more complex measure, is computed by dividing the GNP by a combination of labor and capital, each weighted to reflect its relative contribution to the GNP. The Bureau of Labor Statistics omits agriculture, government, and certain other categories of labor and capital from its calculations. Thus, these published measures reflect productivity in the non-farm, private business sector of the economy.[18]

Available productivity calculations are necessarily estimates and are sometimes criticized on methodological or other grounds. There is general agreement, however, about trends in national productivity. Historical data indicate that total productivity increased at an annual average rate of 1.3 percent from 1889 to 1919. Between 1919 and 1948, the annual rate of increase was 1.8 percent, and from 1948 through the mid-1960s, it accelerated to 2.5 percent per year. Labor productivity during the latter period increased at an annual rate of 3.3 percent. The late 1960s, however, saw the beginnings of a productivity slowdown which today claims national attention. From 1966 to 1977, labor productivity increased at an annual rate of merely 1.6 percent. There has been a continuing retardation of productivity growth since that time and, in 1979, productivity actually stopped increasing and began to decline.

This reversal in the United States' historical productivity pattern has serious implications for a national economy that depends on productivity increases for sustained growth. Declining productivity is, in part, responsible for the inflation and erosion of real wages—a factor that has characterized

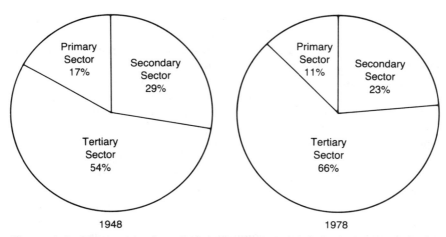

Figure 1-4. Changes in the relative significance of the primary, secondary, and tertiary sectors of the economy are reflected in their contributions to the Gross National Product. The rise of the services sector over the last 30 years has led some investigators to characterize the United States as a "post-industrial society." (Source: Bureau of Labor Statistics.)

the American economy during the past decade. When wage increments exceed productivity growth, prices rise and there is a resulting decline in purchasing power and in the real standard of living. In addition, rising prices have weakened the competitive position of American goods and services in international markets. The United States now has the slowest rate of productivity growth of any non-communist nation. In Japan, by way of contrast, labor productivity is increasing at an estimated six percent per year.[19]

Among other things, productivity can be favorably or adversely affected by changes in the composition of the national workforce. The early to mid-twentieth century period of sustained productivity growth coincided with a movement of workers from relatively non-productive agricultural jobs into manufacturing, where productivity was enhanced by the combined impact of scientific management and technology. Similarly, the recent years of productivity slowdown have coincided with a significant expansion in office-oriented white collar employment, from 21.6 percent of the total workforce in 1950 to slightly more than 50 percent by 1978. Most economists and sociologists view this expansion as a result of a fundamental transition in the American economy. National economies are divided customarily into three sectors: (1) *primary*, which consists of agriculture and the extractive industries such as fishing, mining, and forestry; (2) *secondary*, the manufacturing sector;

and (3) *tertiary*, the services sector. In the United States, the balance of employment among and within those three sectors is shifting in a manner that has important implications for the future of productivity improvement in general and office automation in particular.

1. *The Primary Sector.* While the primary sector still dominates the economies of most of the world's nations, primary sector employment in the United States has declined steadily since the beginning of this century. In 1978, farm employees comprised less than three percent of the total workforce. By 1990, their total will barely exceed two percent.

2. *The Manufacturing Sector.* In terms of its contribution to total employment, the secondary or manufacturing sector dominates the economies of western Europe and Japan and is an important component of the American economy as well. For purposes of this discussion, it is significant to note that, while most of the workforce in the secondary sector consists of machine operators and other blue collar employees who are directly involved with the

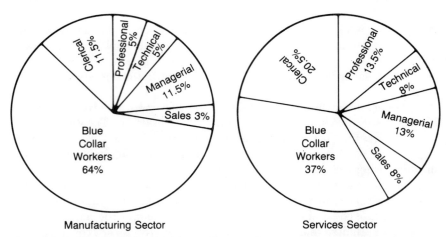

Figure 1-5. The growth of white collar employment has created new opportunities for productivity improvement. The services sector has long been characterized by the predominance of white collar workers. The substantial white collar component within the manufacturing sector is, however, frequently overlooked. (Source: Bureau of Labor Statistics.)

production of manufactured goods, the manufacturing industries do employ a substantial number of non-production workers. Due, in part, to growth in the complexity of business enterprises, the number and percentage of non-production workers in the manufacturing sector has increased steadily since the turn of the century.[20] Today, they total approximately 28 percent of the manufacturing sector workforce—up from 19.1 percent in 1939. In some industries, such as chemicals and publishing, their total exceeds 42 percent of the workforce. Given the current interest in robotics and other innovations in factory automation, the percentage of non-production workers in the manufacturing sector is likely to increase.[21]

3. *The Services Sector.* While there has been a significant increase in white collar employment within the manufacturing sector, the dramatic rise of the tertiary or services sector has contributed most significantly to the growth of the office workforce. The services sector cannot be simply and exclusively equated with office employment.[22] At one extreme, it includes barber shops, moving and storage companies, grocery stores, and other businesses which have a very limited, if any, office component. At the other extreme, however, are some of the most office-oriented, paperwork-intensive

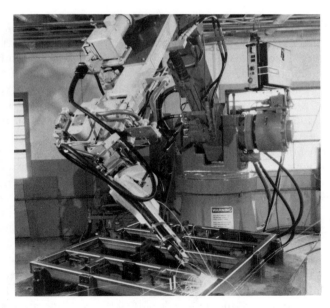

Figure 1-6. Continued productivity in the manufacturing sector depends heavily on technological change such as the use of industrial robots. Here, a robot welds a base assembly for a computer mainframe. (Courtesy: Cincinnati Milacron)

components of the American economy: non-industrial corporations in such fields as banking, insurance, law, and real-estate; education; health care; and government at all levels. The size and increasing importance of the tertiary sector has led economists to characterize the United States as a "post-industrial" society.[23] The services sector now accounts for over 50 percent of the United States' total workforce, and that percentage is expected to increase. Most important in terms of this discussion is the services sector's office orientation. Over 70 percent of its employees work in white collar jobs, and in some segments of the tertiary sector, such as banking and insurance, that percentage is even higher.

Understandably, opportunities for productivity improvement are greatest in those sectors of the economy that utilize the most resources, especially labor. As previously noted, the productivity gains of the early and mid-twentieth century were largely the result of the successful application of technology as a substitute or enhancement for labor in the production of food and manufactured goods, the historically dominant sectors of employment in the United States. But, as described above, during the last several decades, there has been a significant and continuing shift in employment from factories to offices, thereby necessitating a corresponding shift in any proposed productivity improvement efforts. To have an impact on the majority of the workforce, such efforts must be directed at the office.

The Technological Solution

Recognizing the increased significance of the office, work methods analysts, human resources specialists, and others have initiated and implemented a wide range of programs for the improvement of white collar productivity. They point out, for example, the relative youth and inexperience of the office workforce—two attributes which historically have been associated with low productivity. By 1990, persons over 55 years of age will total just 11 percent of the workforce—down from 17.5 percent in 1970.[24] These analysts further note the general lack of commonly accepted productivity measures for white collar employees, especially managerial and professional personnel whose work patterns, as previously discussed, often appear unproductive. They emphasize the importance of clearly defined job descriptions and performance expectations, improved employee motivation, effective supervision, appropriate education, and the careful planning and scheduling of work.[25]

While all of these factors are undoubtedly meaningful, most economists agree that, historically, the most significant increases in productivity have been the result of *technological change*—advances in technology that favorably influence the cost of producing goods and services.[26] Regardless of the application of other approaches to productivity improvement, the technology prevailing at a given time sets limits to the quantity of output which can be produced with a given set of inputs. Technological change, in effect, increases the production function by raising those limits.

Technological change is typically embodied in capital equipment. For example, this was true with regard to the substitution of machinery for labor which led to productivity increases in agriculture in the nineteenth century and in manufacturing in the early to mid-twentieth century. The current productivity slowdown, however, has been accompanied by a similar decline in the amount of capital utilized per hour of labor. From 1948 to 1966, a period of productivity growth, fixed investments in capital stock increased steadily at an average annual rate of 3.1 percent. From 1968 to 1973, the annual growth rate for capital stock dropped to 2.8 percent. Since 1973, it has averaged around 1.7 percent. Within the manufacturing sector, there has been a recent tendency to rely on labor as an alternative to capital investment—a marked departure from previously successful production management strategies.[27] As a result, America's industrial capacity is lagging. An analogous situation prevails in the office where the value of capital equipment per worker has not changed significantly in over 30 years. Despite the history of mechanization discussed earlier in this chapter, the typical office worker is given only about $2,000 worth of capital equipment with which to work. Allowing for intervening inflation, capitalization per office worker is lower now than it was in the mid-1950s.

But the mere enhancement of labor with capital equipment will not insure productivity improvement or its counterpart, cost reduction. While it is possible in many cases to improve labor productivity through additional capitalization, increases in total productivity, as defined at the beginning of this chapter, depend on cost-effective substitutions of capital for labor. Furthermore, some economists contend that the intensified application of prevailing industrial technology—represented, for example, by oil, centralized electrical plants, and the airplane—cannot yield sufficient productivity improvements to support continued economic growth. Instead, new technology, better suited to the national economic changes already discussed, is required.[28] A similar requirement applies in the office where much of the prevailing technology dates from the early industrial period when the white collar workforce was smaller and information processing requirements were less complex. Increased purchases of typewriters, calculators, and copiers—improved as they have been during the last quarter of a century—will not solve the productivity problems of the office in the 1980s and beyond. Effective solutions to those problems will require the technological change embodied in the automated office systems reviewed in the remainder of this chapter and discussed, in detail, in subsequent chapters of this book.

AUTOMATED OFFICE TECHNOLOGIES

Office Information Systems

As noted earlier in this chapter, white collar employees are increasingly characterized as "information workers."[29] The description is appropriately

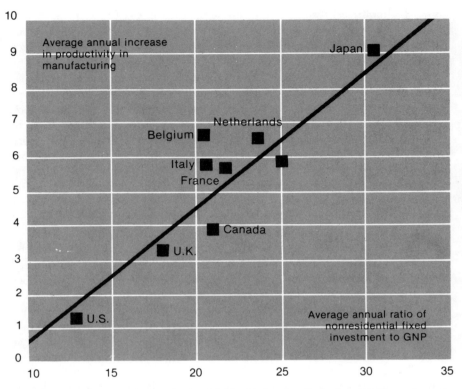

Figure 1-7. Historically, nations with the highest ratios of capital investment to Gross National Product have enjoyed the highest ratios of productivity growth. In the United States, the recent productivity slowdown has been accompanied by a similar decline in the amount of capital utilized per hour of labor. (Reprinted from the *Journal of Micrographics*, July/August 1980. Source: U. S. Bureau of Labor Statistics and Organization for Economic Cooperation and Development.)

applied in several senses. In the broadest sense, most white collar workers rely on information in the form of knowledge gained through education. In the United States, the white collar workforce contains a relatively high percentage of college-educated persons, and continuing education is increasingly viewed as a requisite for effective job performance. The recognized significance of education is not limited to managerial and professional workers, but extends to clerical and technical personnel as well. Apart from formal education, the performance of white collar workers generally is enhanced by information acquired from such sources as: (1) technical books and journals; (2) meetings and publications of professional and trade associations; and (3) newspapers, magazines, and broadcast media.

In a narrower sense, white collar personnel depend heavily on information created within, and pertaining to, their own work environment. This information—which is obtained through meetings, telephone conversations, the examination of files, and related activities—forms the basis for decision-making. Within a corporation or institution, for example, financial officers need information to evalutate capital investments effectively. Engineers need it to make decisions about product development, facilities design, and construction projects. Managers and supervisors need such information to allocate personnel resources and to evaluate the performance of subordinates. Marketing personnel need this material to develop effective sales strategies and to minimize wasted time. While much additional research is required for a clear understanding of the role of information in decision-making, decisions are increasingly viewed as the office's primary "product," and the availability of information is presumed to affect both their quality and quantity. Information continually is regarded as an important corporate or institutional asset and, in the office, information-related activities dominate the workday.

The office itself can be viewed as a system for the creation, processing, storage, retrieval, reproduction, and dissemination of information. In the automated office, new and established technologies enhance or replace labor in the performance of these information-related tasks. The relevant technologies include word processing for the creation of information; computers, micrographics, and other technologies for information, processing, storage and retrieval; reprographics for the reproduction of information; and electronic mail for information dissemination. These technologies are briefly described in the following sections. They are discussed in detail in subsequent chapters.

Word Processing

In the office, correspondence, reports, and other typewritten documents have been the primary vehicles of information communication since the beginning of this century. The cost of creating such documents, as noted at the beginning of this chapter, has risen significantly since that time. Word processing technology and concepts were developed initially to reduce the cost of document creation. When introduced in the early 1970s, the term "word processing" encompassed three product groups: dictation equipment, automatic typewriters and text-editing machines, and copiers. Today, however, conventional copiers are typically excluded from discussions of word processing, although the emergence of the "intelligent copier" indicates the interrelationship between document creation and reproduction. The role of copiers and related reprographic equipment in the automated office is discussed later in this chapter.

Dictation equipment, a comparatively old product group that has improved steadily in both performance and acceptance, offers a convenient and cost-effective means of converting information reflecting office workers' decisions and actions from thoughts into words. In a fully configured word processing system, dictation equipment provides input to automated text-editing equipment that generates typewritten documents. At present, humans, not machines, serve as transcriptionists who are intermediaries in this input process. Future technologies may permit the direct conversion of spoken words into written communications or, as discussed later in this book, voice recordings may selectively replace some written communications.

Automated text-editing systems and related devices, such as memory and intelligent typewriters, often are considered synonymous with word processing. While available individual models may differ in operating methods and capabilities, they are all based on the concept of captured keystrokes. In conventional typing, keystrokes initiate the printing of human-readable characters on paper, but the keystrokes themselves are "lost" in the sense that the operator's keystroking action must be repeated to create multiple typewritten originals or to accommodate document revisions or format changes. With automated text-editing systems and related products, however, the operator's keystrokes are captured in machine-readable form on magnetic recording media from which they can be printed, revised, or manipulated. In certain types of applications, especially those involving extensive revisions or retyping, the manipulation of captured keystrokes can increase productivity significantly and reduce the cost of document creation. Apart from document creation, the capturing of keystrokes is indispensable to other aspects of office automation, some of which have nothing to do with paper documents. Once recorded in machine-readable form on magnetic media, the information content of office communications can be inputted, for example, to computer-based storage and retrieval systems, recorded on microfilm or microfiche, or transmitted electronically to remote locations.

Computers in the Office

Through the late 1970s, the implementation of word processing systems, especially the selection and procurement of text-editing equipment, dominated the efforts of administrative support specialists, information systems analysts, and others responsible for the development of automated office systems. However, considerable recent attention has been given to the role of computers as a means of dealing with the office's information processing problems and of integrating otherwise incompatible technologies. Computer manufacturers and systems designers are taking an increased interest in office applications. As a result, computer technology likely will prove to be the focal point of office automation in the 1980s.

As discussed earlier in this chapter, office-oriented computer applications have been commonplace since the 1960s. Through the early 1970s, however, the office's data processing requirements were typically met by centralized computing facilities. Data generated or acquired by individual offices were submitted, at predetermined intervals, to a computing center for keypunching and batch processing. The results were returned to the originating offices in the form of printed reports. While batch-processing and centralized computing remain in wide use, computers and peripheral equipment frequently are encountered in the office itself. By the mid-1970s, online terminals and time-sharing operating systems made local interaction with remote computers conveniently possible. Today, printing and display terminals are encountered widely in office applications, and improvements in direct-access storage devices and data base technology have contributed to the development and increasing utilization of online information processing, storage, and retrieval systems. Equipped with a terminal, an office worker now can have rapid, convenient, and cost-effective access to internally developed management information systems, as well as to a wide range of publicly available online information resources. Improvements and enhancements in terminal features, such as the ability to display data graphically and the development of touch-sensitive screens as an alternative to typing, will facilitate their acceptance as a management tool. Apart from their use for online information storage and retrieval, computers play an important role in certain types of micrographics retrieval systems, electronic mail and message systems, and word processing applications.

While online, time-shared access to remote computers is an important and convenient feature in the automated office, recent improvements in the manufacturing of electronic circuitry—specifically, the manufacturing technique called large-scale integration (LSI)—have made it economically possible for an office to install and operate a computer dedicated to its own information processing requirements. The first of these dedicated computers—the *minicomputers*—appeared in offices in the early-1970s and quickly gained popularity. Smaller and less expensive than their large-scale counterparts, these minicomputers allowed an organization to decentralize or "distribute" its data processing resources, thus gaining flexibility and improving cost-effectiveness. *Microcomputers,* a product of the late 1970s, represent the latest extension of this trend toward locally pervasive computing power. Designed specifically for the office environment, a microcomputer features a low-cost microprocessor as its central processing unit. Microcomputers employ comparatively inexpensive peripherals, pre-written software, and simplified user-programming for cost-effective information processing in office applications that could not have previously justified a dedicated computer or conveniently obtained time-shared access to a remote computing facility. A special class of microcomputers, *intelligent terminals,* can access other computers via telecommunications, thus permitting their integration in information-processing networks.

Micrographics

While text-editing and computers are based on relatively new technological developments, micrographics is an example of an established technology that has assumed new importance and popularity in the automated office. Microforms are, of course, well suited to the "paperless" office concept emphasized by some information management specialists and mentioned earlier in this chapter as one of the characteristics which distinguishes current office technology from its predecessors. From the standpoint of personnel cost reduction and labor productivity improvement, a well-designed micrographics system can eliminate clerical labor and facilitate rapid retrieval. In terms of total productivity improvement, microforms conserve space—an important and increasingly expensive capital resource relating to the input aspect of productivity.

Used since the 1920s for the long-term storage of inactive business records, micrographics began to be widely applied to active records in the 1960s. During the period from 1960 to 1970, for example, insurance companies made extensive use of microfilm jackets for active policy files; engineering departments utilized aperture cards for the simplified storage and retrieval of drawings; and government agencies disseminated newly published scientific and technical reports on microfiche rather than in paper form. Since 1970, the micrographics industry has emphasized the development of products for active information management applications and has actively promoted the integration of micrographics and computer technologies. To facilitate acceptance in applications involving active rather than archival information, micrographics equipment has become progressively more versatile, attractive, and reliable. An increasing number of source document cameras, for example, now incorporate microprocessors as controllers. Several cameras feature convenient, self-contained film processors. Readers have improved considerably in quality, cost/performance characteristics, and attention to human factors. Manufacturers are increasingly directing their marketing efforts toward small office microfilm (SOM) applications—an historically neglected and technically inexperienced user community which requires uncomplicated, cost-effective solutions to its information-processing problems. To extend the range of micrographics to applications involving large numbers of growing files, several updatable microfiche systems employ innovative recording technologies.

In terms of the integration of micrographics and computer technologies, computer-output microfilm (COM) has been recognized as a cost-effective alternative to paper printouts since the early 1970s. Acceptance of COM in business applications has increased significantly since that time. Computer-assisted retrieval (CAR), another form of micrographics/computer integration, combines a data base of document images on microfilm or microfiche with a computer-maintained index which is searchable from an online terminal. Compared to conventional computer-based information storage and retrieval systems, CAR offers simplified input, storage economies, and the ability to accommodate photographs and certain types of documents which may not be represented conveniently in machine-readable form.

Reprographics

As a field of information management, reprographics is broadly comprised of those technologies that make reproductions of documents. Technically, micrographics can be considered a subfield of reprographics but, for purposes of this discussion, the scope of reprographics will be limited to technologies which make full-size, eye-legible document reproductions in paper form.

The most commonly encountered reprographics equipment—copiers, duplicators, and typesetters—have been in office use for several decades and are sometimes considered "mature" technologies with little potential for future growth. Because they produce paper documents, some information management specialists view these traditional reprographics devices as outside the current mainstream of office automation with its emphasis on paperless information systems. However, automated or not, paper remains an important information carrier in the office, and several recent product developments suggest the continuing interrelationship between reprographics and other aspects of office automation, especially computer technology and word processing. Copiers and duplicators, for example, are increasingly incorporating integrated electronic circuitry and are utilizing microprocessors as controllers. The result has been a more versatile and reliable machine with excellent cost/performance characteristics when compared to older models. In a few of the newest convenience copiers, the conventional lens and light source have been replaced with fiber optics, producing a compact, less expensive and less complex device. To increase labor productivity, copiers and duplicators can support automatic sorters, feeders, and other peripheral attachments.

In terms of the integration of reprographics and other types of automated office equipment, typesetters can interface with both text-editing and computer systems to simplify the preparation of composed output. Most recent attention has, however, been given to a new type of reprographics equipment—the *"intelligent copier."* While its exact definition remains uncertain, the intelligent copier is a device which can produce multiple-copy paper copies from either paper documents or machine-readable text prepared by a computer or text-editing system. Because such text can be transmitted to it electronically, the intelligent copier is increasingly viewed as an integrative device capable of linking or emulating word processing, computer, and electronic mail systems.

Electronic Communication

In the broadest sense, the field of electronic communications encompasses transmissions via voice, data, and image. In terms of automated office applications, however, most of the interest in electronic communication has centered on *electronic mail*—that group of technologies and equipment designed for the rapid transmission of messages to remote locations.

Telegraphy, developed in the mid-nineteenth century, was the first technology to deliver written messages at speeds faster than those obtainable through conventional postal service or other forms of physical document delivery. TWX and telex, two telegraph-derivative technologies, have been in office use for many years, as has facsimile, the technology which transmits copies of documents. The use of TWX, telex, and facsimile has been limited traditionally to time-sensitive communications. Until the mid-1970s, the cost/performance characteristics of electronic message transmission were limited by a network of telephone and telegraph facilities developed at the beginning of the twentieth century and designed primarily for voice rather than high-speed data or image transmission. The late 1970s, however, saw the beginnings of major changes in this existing communications infrastructure. Encouraged by a decrease of regulatory restrictions and by new developments in communications technologies, these changes are reflected in: (1) the growth of specialized communications carriers for data and facsimile transmission; (2) the development of domestic and international communications facilities; and (3) the use of coaxial cable for intra-building communication linkages. The result has been renewed interest in, and enhancements of, established message transmission technologies, accompanied by the introduction of new products and services designed to address a broader range of office communication requirements.

TWX and telex have expanded to include cross-network communication capability, the ability to store messages for forwarding at a later specified time, multiple addressing services, and compatibility with Mailgram and related message delivery services. Text-editing equipment can be optionally equipped with message transmission capability, allowing it to address compatible text-editing systems or, in some cases, TWX or telex terminals. Facsimile technology has improved considerably. Newer facsimile equipment offers better quality and faster, more reliable service than the equipment offered to offices in the 1960s and early 1970s.

While the technologies discussed above address the problem of long-distance message transmission, computer-based message systems are designed to be used by managerial and professional personnel as a replacement for time-wasting telephone calls and brief internal memoranda. Using a computer as an intermediary between online terminals, these systems employ special software which accepts, routes, and later stores messages. The most complex computer-based message systems feature text-editing capabilities as well as the ability to file documents electronically for later reference.

Finally, information management specialists recently have demonstrated considerable interest in video-oriented technologies, including Picturephone, video conferencing, and video-based image transmission involving paper, microform, or machine-readable files. These and the other technologies described in this introduction are discussed in detail in the following chapters.

2
Word Processing

Dictation systems • Automated text editing • Text editing and related technologies

As introduced in the preceding chapter, the phrase *word processing* intentionally conveys the impression that conventional data processing has emphasized numbers to the exclusion of the textual information that is pervasive in office applications. In the broadest sense, the phrase denotes a group of concepts, technologies, and techniques designed to simplify the recording and communication of the textual information—correspondence, memoranda, reports, and similar documents—essential to office operations. In keeping with this broad definition, a fully configured word processing system includes equipment for the creation, storage, retrieval, and dissemination of textual information. Typically, however, the phrase is used in a narrower sense to denote the application of technology to the creation of typewritten communications. This narrower definition reflects word processing's origins in the German term *textverarbeitung* (literally, the "making of text"), but fails to convey the increasingly multi-faceted operation of the newest word processing equipment. Using this narrower definition as a starting point, however, this chapter describes two groups of word processing products—dictation systems and automated text-editing systems, emphasizing their application to the most commonly encountered office documents. The integration of word processing and other automated office systems will be discussed in subsequent chapters.[1]

DICTATION SYSTEMS

Basic Concepts

Word processing begins with the conversion of the thoughts of a person called a *word originator* to a form suitable for transcription by a typist. The

21

word originator may, for example, be an executive preparing correspondence or memoranda, an attorney writing a brief, a scientist or engineer preparing a technical report, or a physician recording a patient's diagnosis. In each case, the required conversion of thoughts to words customarily is accomplished by: (1) writing them in longhand; (2) dictating them to stenographer; or (3) dictating them to a voice recording machine. While each has advantages, the three methods differ significantly in the type and amount of labor involved:

1. *Longhand Writing.* With regard to longhand writing, special equipment, training, and working conditions are unnecessary. Longhand writing can be performed almost anywhere—in an office, at home, or while traveling. Secretarial assistance is not required. The longhand method is especially well suited to documents containing highly technical terminology that is understandable only to the word originator. The transcribing typist should experience little difficulty with unusual spellings or punctuation, assuming that there is legible handwriting. Longhand writing is likewise appropriate to the composition of multi-page reports and similar documents where reference to previously written paragraphs is essential.

2. *Dictation to a Stenographer.* Although secretarial involvement is not required, longhand writing is a slow process which can consume much of the word originator's time. Since typical writing speeds seldom exceed 10 to 15 words per minute, a longhand draft of a 250-word memorandum will require between 15 and 25 minutes to create. Because the salary of a word originator is typically much higher than that of clerical support personnel, the cost of document creation can often be reduced by having the word originator dictate to a stenographer. When the word originator is properly prepared, shorthand dictation can be two to three times faster than longhand writing. Assuming a dictation rate of 30 words per minute, a 250-word memorandum can be converted from thought to a form ready for typing in as little as 8 minutes. Even though secretarial involvement is required, the resulting reduction in word originator time will, in most cases, offset the additional clerical costs. In addition, immediate interaction between the word originator and the stenographer can clarify ambiguous terminology, spellings, or punctuation, thus simplifying and reducing the cost of eventual transcription. Typically, the typing is performed by the stenographer to whom the communication was dictated, since many stenographers experience difficulty in deciphering shorthand written by others. Some flexibility in the assignment of work must, consequently, be relinquished.

3. *Machine Dictation.* This flexibility can be retained and, in many cases, costs may be reduced further through the application of voice recording equipment. Because dictation to a machine typically can occur at a faster rate than dictation to a stenographer, the word originator's involvement in document creation is minimized. Assuming a speaking rate of 45 to 60 words per minute, a 250-word memorandum can be converted from thought

to a voice recording in less than 5 minutes, with a resulting beneficial impact on document creation costs. As already noted, the use of dictating equipment offers the word originator increased flexibility as well. A stenographer's presence, for example, is not required—an important consideration in offices having a low clerical-to-administrative employee ratio. Given the availability of portable recorders and the other special equipment described later in this chapter, dictation can be performed at home or while traveling.[2]

The potential for improved work effectiveness is, however, not limited to the word originator. The absence of stenographic involvement in dictation results in a significant reduction in the clerical costs associated with document creation. From the standpoint of personnel management, training in shorthand need not be a requirement for clerical employment or job advancement. Additional savings consequently may result from the elimination of special personnel classifications based on now-superfluous clerical skills. From the standpoint of clerical productivity, typing or other work need no longer be interrupted to take shorthand.

While these potential advantages are considerable, the implementation of effective dictation systems requires careful attention to equipment selection and user orientation. These facets of dictation systems design are discussed in the following sections.

Desk-Top Systems

In its simplest form, a dictation system consists of two hardware components:

1. a *recorder* with a microphone designed to capture spoken words as an analog signal on magnetic media; and
2. a *transcriber* with a speaker designed to convert the recorded signal to audible sounds for playback through a speaker or earphones.

The typical recorder is a desk-top device designed to accept an audio cassette or other magnetic medium with capacity for 30 to 120 minutes of voice recording. All available recorders offer some playback capability which allows the word originator to review previously recorded material. When dictation is completed, the cassette must be physically removed from the recorder and taken to a typist who inserts it into a separate transcriber. Although some recorders are true dual-purpose devices suitable for both recording and transcription, the typical transcriber is a less expensive single-purpose unit designed to accept and play back information from previously recorded media.

In addition to the basic operations of recording, transcribing, and tape movement familiar to users of mass-market cassette recorders, dictation systems designed for office use incorporate standard or optional features designed to facilitate the work of word originators and typists. From the standpoint of human engineering, considerable attention has been given to

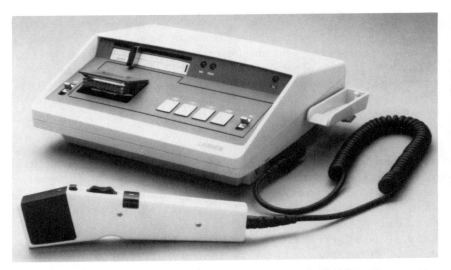

Figure 2-1. The Lanier OMNI-Q is a compact desk-top dictation system which utilizes microcassettes as the recording medium. The word originator can electronically record a special tone followed by instructions that alert the transcriptionist to any changes to be made in previously dictated communications, eliminating time-wasting re-typing. (Courtesy: Lanier Business Products, Inc.)

the nature and placement of operator controls. Most recorder microphones, for example, are designed to be held and operated with one hand. Transcribers usually are equipped with an orthopedically designed foot pedal. Variable speed controls allow the transcriptionist to adjust the playback rate to typing speed with a minimum of voice distortion.[3] Following pauses for typing, many transcribers automatically will repeat a few previously played words before playing new material. Several newer recorders feature voice-operated relay (VOR) technology through which the recording mechanism is activated only when the word originator is actually speaking, thus facilitating transcription by eliminating tape gaps resulting from periods of thought or other pauses.[4] Several techniques, including paper index strips and electronic coding, can be used by the word originator to mark the locations of special instructions or priority communications on tape. As is true of other types of information processing equipment, recorders and transcribers are increasingly microprocessor-controlled for improved versatility and reliability.

At the time this chapter was written, prices for desk-top recorders and transcribers were in the $150 to $400 range per unit. While the more expensive devices offer a wider variety of operating capabilities, a combination of one desk-top recorder and one transcriber with features satisfactory for most applications can be purchased for around $400. Cost-justification depends on the attainment of a volume of dictation activity sufficient to

Figure 2-2. The Dictamation DCX III is a microprocessor-controlled desk-top dictation recorder/transcriber which also can function as a telephone answering machine. A compatible portable recorder is available for use with the system. (Courtesy: Dictaphone Corporation)

permit the recovery of equipment costs through savings in word originator and clerical time. In most applications, the volume of required activity is relatively modest. As an example, assume that the purchase price of a desk-top recorder and compatible transcriber (a total of $400) is amortized over a useful equipment life of 36 months, for an equivalent monthly cost of approximately $11.00. If each use of the equipment reduces the cost of creating a 250-word document by $2.00 when compared to dictation to a stenographer, then the system is cost-justifiable at 6 such uses per month. When compared to longhand writing, each use of the system reduces the cost of creating a 250-word document by about $3.00, and the equipment is cost-justifiable at 4 uses per month.

Portable Recorders

While the combination of a desk-top recorder with a simple, separate transcriber has proven extremely popular, certain applications are better served by alternative equipment configurations. The potential utility of portable recorders was introduced briefly at the beginning of this chapter. The typical portable recorder is an attractive compact device with many features found on desk-top units. Special mini- and micro-cassettes, discussed later in this chapter, are often used to decrease recorder size, although some portable units will accept the standard cassette. Most available units can be held easily and operated with one hand; they are small and light enough to be unobtrusively carried. While they can be used in an office environment, portable recorders are designed specifically for the preparation of correspondence, memoranda, and similar communications at home, while travelling, or in other situations where a desk-top recorder is unavailable. The resulting recorded media must, of course, be taken to a compatible transcriber—typically, one of the desk-top units described in the preceding section. If the word originator is travelling, recorded media can be mailed to the office or transcribed by locally available secretarial services. The centralized dictation systems described in the next section of this chapter will sometimes accept communications from portable recorders transmitted via the telephone. Many word originators also find portable recorders useful as an alternative to conventional note-taking in such applications as field inspections, inventory preparation, and laboratory experimentation.

Prices for portable recorders begin at around $200. The most expensive units, which can cost several times that amount, frequently offer a combination of operating and cosmetic features. Cost-justification is determined in the manner described in the preceding discussion of desk-top units. The purchase price of the system is amortized over a reasonable useful

Figure 2-3. The availability of portable recorders lends flexibility to dictation systems. Sizes vary. From the left, the recorders shown here will accept a minicassette, a standard cassette and a microcassette. (Courtesy: Dictaphone)

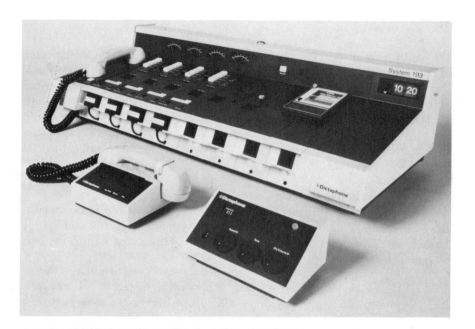

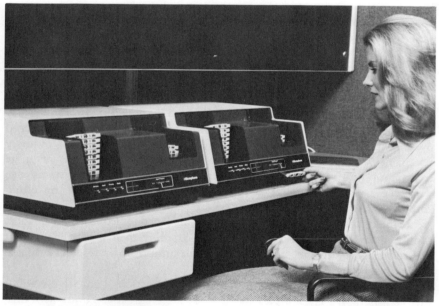

Figure 2-4. In centralized dictation systems, the stand-alone desk-top recorder is replaced by a simple, inexpensive microphone or an ordinary telephone wired to a remote recorder. Some "tank-type" systems feature a continuous, non-removable loop of magnetic tape designed for many hours of simultaneous recording and transcription. Others are designed for the automatic feeding of individual tape cassettes. (Courtesy: Dictaphone Corporation)

life, and the resulting monthly equivalent cost is divided by the dollar value of the savings anticipated each time the system is used. Unlike desk-top systems, however, portable recorders are an alternative only to longhand writing, since a stenographer is presumably unavailable in those work situations where the portable recorder is utilized. As is true of desk-top units, portable recorders are important automated office products because they increase managerial productivity and have a correspondingly beneficial impact on office costs.

Centralized Systems

As noted above, media containing communications recorded by desk-top or portable units must be physically removed and taken to a compatible transcriber. This poses two problems for word originators with occasional or sporadic dictation requirements:
1. Recording devices must be provided for each word originator. While the equivalent monthly cost, as previously discussed, is relatively low, justification may prove difficult or impossible to demonstrate in cases where the equipment rarely will be utilized.
2. To avoid repeated trips to the transcribing station, short communications are typically accumulated until a cassette or other medium is nearly full. This results in both an uneven flow of transcription work and comparatively long turnaround times for material at the beginning of a cassette.

These problems can be minimized or eliminated with a centralized system in which one or more remote recorder/transcribers are used by several or many word originators. The design of such systems differs significantly from those previously discussed. The familiar desk-top recorder is replaced by a simple, inexpensive microphone or, in some cases, by an ordinary telephone wired to a centrally located recorder to which one or more transcriptionists are likewise connected by earphones. Some systems also will accept calls from external telephones, a useful feature for travelling employees. As new material is dictated, previously recorded communications are being transcribed. Word originators are encouraged to dictate short communications immediately, rather than batching their work. The regular flow of input to the recorder facilitates maximum utilization of available transcriptionists.

The capabilities and capacities of available centralized dictation systems vary considerably. The simplest systems are merely desk-top units with dual cassette drives. While one tape is being recorded, the other is being transcribed. Such systems may serve a half-dozen word originators but cannot support simultaneous dictation. Systems designed for larger applications employ automatic cassette changers to obtain increased recording time. In some cases, discrete media are replaced by a continuous, non-removeable loop of magnetic tape designed for simultaneous recording and transcription. Such "tank-type" systems can accommodate up to 12 hours of recording and will support simultaneous input from several word originators. The most advanced centralized systems employ microprocessors and cathode ray tube (CRT) displays to: (1) monitor and communicate work status information to supervisors; (2) expedite the transcription of priority communications; (3) ensure the equitable allocation of work to individual typists; (4) measure productivity; and (5) predict job turnaround.

Figure 2-5. The most advanced centralized dictation systems use microprocessors and display terminals to monitor and communicate work status information to supervisors. With the Lanier Super-Vision III, a word processing center supervisor can utilize a CRT display to control workflow and generate management reports and work performance summaries. (Courtesy: Lanier Business Products, Inc.)

Prices for centralized dictation systems vary with the particular configuration and the number of microphones supported. Compared to a configuration employing multiple recorders and transcribers, the total cost of a centralized system usually will prove lower where many word originators must be supported, although individual dictating activity may be low.

Recording Media

As noted in the preceding sections of this chapter, the medium employed in most desk-top and portable recorders, as well as in some centralized systems, is a voice-grade magnetic tape packaged in a double-core plastic container called a *cassette.* The standard or Philips-type cassette measures 2.5 by 3.86 inches and is widely available from office suppliers and various retail outlets. It permits 30 to 120 minutes of recording and can be played on a wide range of transcribers, including most of the cassette recorder/players in consumer use. Two smaller variants of the standard cassette, the *minicassette* and the *microcassette,* are designed primarily for use in portable recorders where they facilitate compact equipment design. Adapters are available to permit certain transcribers to accept cassettes of various sizes.

IBM, whose earlier dictation equipment employed magnetic-coated belts rather than cassettes, now utilizes special magnetic-coated disks in its System 6:5. Each disk contains up to 6 minutes of recording—the time required for a typical 1- or 2-page business communication. For purposes of transcription, individual disks are stacked in cartridges, each of which has a total capacity of 5 hours of recorded material.

Cassettes and disks are examples of discrete recording media, so called because they can be removed from a given recorder or transcriber and used on others as necessary. As noted in the preceding section, some centralized dictation systems employ a "captive" recording medium consisting of an endless loop of magnetic tape which is continuously cycled through a recorder/transcriber. The recording capacity of the largest available endless loop system is approximately 12 hours.

User Orientation

Although the concept of voice recording is an old one, the performance capabilities of dictation systems have improved steadily through the past several years. Attractive new features have been regularly introduced, and there is every indication that the trend toward greater performance will continue. In the office environment, dictation systems recently have enjoyed significantly increased user acceptance. However, while they offer potential for significant improvements in productivity, dictation systems do require changes in the word originator's work orientation. These adjustments can, however, be facilitated by a brief, well-designed orientation and instruction program. This program must emphasize those aspects of effective dictation that require a modification of established work habits. The word originator must be taught, for example, to: (1) dictate from an outline or brief notes; (2) spell unusual terms or unfamiliar proper names; and (3) include any special instructions concerning the type of paper or other typing medium to be used, the number of copies to be prepared, and special formats to be observed. Most dictation systems manufacturers have developed special training aids, in the form of booklets and audio-visual instruction packages, which can be supplemented with locally prepared materials or brief orientation sessions.

AUTOMATED TEXT EDITING

The Problem of Conventional Typing

As described in the preceding sections of this chapter, dictation systems facilitate the conversion of a word originator's thoughts to a form suitable for transcription. Until voice-driven printers become widely used, the transcription process will continue to require typewriting or a similar form of keystroking.[5] As discussed in Chapter One, the typewriter represents one of the earliest examples of the application of automation to office work, and the typewritten document remains the primary vehicle of business communication. Yet, as important as it is, work methods analysts and information management specialists have long recognized that diminished productivity, and consequently increased expense, are the inevitable result of three shortcomings inherent in conventional typing:

1. Repetitive text—such as the information content of form letters and mass mailings, the introductory sections of proposals and reports, the standard paragraphs of contracts, and similar "boilerplate" material—must be individually retyped to produce multiple typewritten originals or to be otherwise reused. Such retyping is not only time consuming, but involves considerable potential for error.
2. Whether single or multiple originals are involved, keystroking errors are inevitable. When they occur, the usual methods of correction (erasure, overlaying with opaque material, or starting over) interrupt keystroking, take considerable time, and can deface documents.
3. Entire documents must often be retyped to accommodate even minor revisions and/or format changes. Some documents—such as technical reports, proposals, and legal briefs—routinely require several retypings to accommodate revisions ranging from content changes to rearrangement of the order of sentences or paragraphs. Each retyping introduces the potential for keystroking errors in those portions of the document which were typed correctly in previous drafts.

These shortcomings inevitably result in degraded work, decreased clerical productivity, and increased document production costs. As work methods analysts point out, a clerk who ostensibly types 60 words per minute actually may achieve work equivalent to less than 25 percent of that rate when all retyping and error correction routines are considered.

Historically, the indicated problems of conventional typing have been addressed in several ways. Through the early 1960s, automatic typewriters—such as the Friden Flexowriter and American Autotypist—simplified the task of repetitive typing by pre-recording keystrokes on punched paper tape or similar media that could be played back through a typewriter to create multiple originals. The error correction and document revision capabilities of such devices were, however, quite limited. Computer-based editing software became widely available on a time-sharing basis in the mid-1960s, but stringent editing rules and elaborate coding requirements made such software more suitable for the entry and editing of programs and data in online files than for the creation of typewritten office documents. Automated text-editing as a practical office procedure utilizing specially designed equipment dates from 1964 when IBM introduced the Magnetic Tape/Selectric® Typewriter (MT/ST), the first device capable of simultaneously recording keystrokes on paper and on an easily changeable magnetic tape. Besides automatically typing repetitive documents, the MT/ST allowed typists to easily correct keystroking errors and make changes in previously typed text without extensive retyping. Although it now seems unsophisticated and inconvenient when compared to recently introduced equipment, the MT/ST was a trend-setter for the future use of magnetic media in automated text-editing systems.

The next sections of this chapter survey the current state of the art in text-editing equipment. Because equipment enhancements and special features are being introduced at a rapid pace, the discussion emphasizes essential system characteristics rather than the capabilities of specific devices.

Typewriter-Based Systems

As with other types of information processing equipment, basic text-editing hardware configurations consist of input and output devices, a storage medium, and a control unit or central processor. Text entered at the keyboard of the input device is captured by the control unit and recorded on magnetic media (a card, tape, or disk) for later correction, revision, or playback by the output device. This basic configuration is available in several variations. The oldest and least expensive combines the input and output devices in a single typewriter.

Through the mid-1970s, most such combined systems employed the familiar IBM Selectric Typewriter in either the office or heavy-duty model. The simplest systems—some of which remain in use—merely attached a control unit to the customer's own Selectric, using a special baseplate. While this approach offered text-editing capabilities at a relatively low cost, most manufacturers preferred to make one or more modifications to a heavy-duty Selectric (the model used as a component in computer terminals) before offering it as part of a pre-configured system. This is, for example, the approach evident in the IBM Magnetic Card Selectric Typewriter (MC/ST), which still comprises the majority of installed text-editing devices. Along with high output quality, the convenience and versatility of its interchangeable typing elements are the obvious attractions of the Selectric and account for its popularity with clerical personnel. By using it as the

Figure 2-6. Typewriter-based word processing systems, such as the IBM Mag Card II, combine the input and output device in a single typewriter—in this case, an IBM Selectric. While display-oriented text-editing systems account for the vast majority of new equipment sales, the installed base of text-editing systems still includes the many typewriter-based devices manufactured and sold during the 1970s. (Courtesy: IBM Office Products Division)

input/output device in their configurations, manufacturers of early text-editing systems were able to successfully overcome resistance to change.

Yet, despite its attractiveness and versatility, the Selectric is slower than the typewriter-quality print mechanisms used in recently developed computer terminals. Even when operating at its maximum operating speed of 180 words per minute, the Selectric can result in significant wasted operator time in applications involving substantial amounts of printing. In 1975, Xerox introduced its 800 Series Electronic Typing Systems which, while resembling the IBM Magnetic Card typewriters in editing capabilities, employed the Diablo Hy-Type printer as a faster alternative to the Selectric® Typewriter. The Diablo printer featured an interchangeable rimless printwheel on which characters are represented by embossed metal slugs at the ends of spokes. A similar mechanism is used in printers manufactured by Qume Corporation. A slightly different approach is taken by the NEC Spinwriter, which features interchangeable thimble-shaped printing elements. These alternative printers operate at speeds of 350 to 550 words per

Figure 2-7. Many word processing systems include printers with interchangeable printwheels on which characters are represented by embossed metal slugs at the ends of spokes. The printwheels may be either metal or plastic and are available in the most popular type styles. (Courtesy: Daisytek, Inc.)

minute and offer output quality and versatility comparable to that of a Selectric® Typewriter.[6] While used in several typewriter-based text-editing systems, these printers most often are encountered in the teleprinter devices discussed in Chapter Three and in display-oriented text-editing systems.

Display-Oriented Systems

Regardless of the speed of their print mechanisms, typewriter-based text-editing systems have one obvious disadvantage: their keyboards are not available for input while text is being printed. This disadvantage is addressed by the rapidly increasing number of text-editing systems which feature a cathode-ray tube (CRT) as the input device and a separate typewriter or other printer for output. Information entered at the keyboard is displayed for editing on the CRT screen. When all desired corrections and revisions have been completed, the displayed information is dispatched to the printer, leaving the keyboard again free for input. The use of continuous paper stock or automatic sheet feeders permits unattended printer operation.

While display-oriented text-editing systems bear an obvious resemblance to computer terminal devices, the operation is designed intentionally to emulate ordinary typewriting, thereby minimizing operator training requirements. With the exception of special function keys unique to particular models, the input keyboard is identical with that of a Selectric

Typewriter. Much recent attention has been given to the development of letter-size or larger screens which display an image resembling a typewritten document, thus enabling the operator to visualize completed pages prior to printing. While such full-page screens may display 60 or more lines of text, some systems provide a smaller screen which presents 25 lines or less at one time. Although such systems display only a portion of a document, a capability called *scrolling* allows the operator to examine those lines which precede or follow those on the screen.[7]

Regardless of screen size, the output device is typically a Qume or Diablo printer, although innovative printers are attracting increased market attention. The IBM Office System/6, for example, can be configured with an ink-jet printer which sprays ink droplets to form characters at speeds of up to 920 words per minute. Other systems can operate with line printers or even faster xerographic printers and other "intelligent copiers" (discussed in a later chapter). For applications involving the production of lengthy documents, several systems can be configured with a printer capable of accepting paper up to 18 inches wide. For the production of scientific and technical documents requiring the integration of text and mathematical or other symbols, Qume and Diablo have developed printers with two print mechanisms—one designed to hold a conventional alphanumeric printwheel, the other a printwheel with the required special symbols. In an alternative approach, the NEC Spinwriter combines conventional alphanumerics with mathematical symbols on a single print thimble.

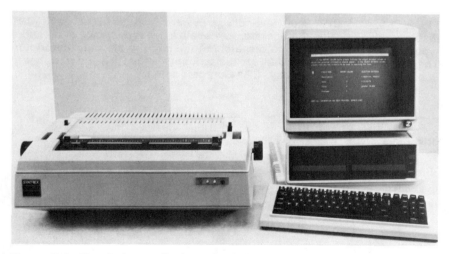

Figure 2-8. Stand-alone, display-oriented text-editing systems are microprocessor-controlled devices which provide a CRT display with keyboard for input and a separate typewriter or other printer for output. The Syntrex Aquarius workstation features a partial page CRT mounted atop a cabinet containing dual mini-floppy disk drives. (Courtesy: Syntex, Inc.)

Figure 2-9. The CPT 8000 text-editing system features a full-page CRT display which enables the operator to visualize the appearance of an entire typed sheet of paper prior to dispatching the page to an attached printer. The screen is housed with dual standard floppy disk drives and a keyboard in a single chassis. (Courtesy: CPT Corporation)

In 1975, the IBM Magnetic Card Typewriter and other typewriter-oriented systems dominated the text-editing market, and display-oriented systems accounted for a mere one percent of total installations, with only three manufacturers represented. By 1978, however, market researchers estimated that display-oriented devices comprised 20 percent of the then installed base of 265,000 text-editing systems. Today, over 20 vendors offer such systems. Taken together, they account for the vast majority of new text-editing installations and are expected to continue to dominate the market, as experienced word processing users—in the manner of their data processing counterparts—upgrade their typewriter-based text-editors to more complex systems. At the time this chapter was written, typical display-oriented system prices were in the $15,000 to $20,000 range, although several models priced at less than $10,000 have been announced recently.

By way of contrast, the price range for typewriter-based systems is typically $5,000 to $7,000. Proponents of display-oriented systems point out, however, that their higher equipment costs can be justified through productivity improvements inherent in simultaneous keyboarding and printing.

One-Line Displays

The systems described above feature a CRT screen capable of displaying a full or, in some cases, a partial page of text. The *one-line display* text-editing system is a recently introduced variant of CRT-based text-editors and combines a typewriter-based system with a calculator-like light-emitting-diode (LED) or gas plasma display featuring 15 to 40 characters of text accompanied by coded operator instructions. This single line essentially displays the contents of a small internal memory in which keystrokes are captured and edited prior to printing them onto paper. A line of text is entered at the keyboard and displayed for error correction or revision. Characters can be quickly and easily changed by a combination of backspacing

Figure 2-10. Equipped with a one-line display which serves as a "window into memory," the Adler SE-2000D allows the operator to correct errors before they are printed on paper. (Courtesy: Adler-Royal Business Machines, Inc.)

and overstriking. When all changes have been made, the typewriter mechanism prints the line. While keystroked characters also may be recorded on magnetic media for later recall and revision, the one-line display primarily is designed to facilitate the error-free initial production of correspondence, memoranda, and other comparatively short documents. As such, it competes with the IBM Correcting Selectric and equivalent typewriters, as well as the memory and intelligent typewriters discussed later in this chapter.

Dedicated vs. Shared Processors

Regardless of the input/output configuration selected, the text-editing system has a central processor or control unit that has two basic functions:
1. To record text entered via keystrokes in machine-readable code on magnetic storage media; and
2. To interpret the code and display or print the text, in human-readable form, on paper or on a screen.

As such, the text-editing system can be compared in design and purpose to the central processing unit (CPU) in computer systems. Text-editing system processors typically are divided into two groups: *dedicated* and *shared*. This division is nominally based on the ratio of input and output devices to the central processor, but there are significant performance differences as well. A dedicated processor, as the name implies, controls one input and one output device. Commonly described as *stand-alone* systems, they dominate the text-editing market. While the IBM Magnetic Card Typewriters and other early stand-alone units had hard-wired controllers with very small internal memories and fixed logical circuitry, most newer systems are controlled by microprocessors whose editing and other capabilities are embodied in programs stored in specially designed electronic circuits called Read-Only Memories (ROMs). These resident programs, which are pre-written by the equipment manufacturer, are called *firmware* to distinguish them from conventional programs or *software*, which are stored on magnetic media and entered into the system's internal memory prior to use. A small, but increasing, number of stand-alone text-editing systems are, however, software-based. The software approach, in which pre-written editing and other operating programs are supplied on floppy disks by the equipment manufacturer, is attractive because it facilitates the addition or modification of editing capabilities without alterations in the machine itself. Only the program disk need be replaced. Upgrading of firmware-oriented machines requires a replacement of internal memory components. While easily accomplished, this replacement usually must be performed by a trained service technician. The next chapter—which deals, in part, with microcomputers—presents a more detailed discussion of microprocessors as controllers. Stand-alone text-editors can be viewed as microcomputers designed for a specific application—automated production of typewritten documents.

Shared processor (often called *shared logic*) systems are actually time-shared computers designed to support multiple input and output devices simultaneously. The processor itself is typically a minicomputer which has greater power than its stand-alone counterpart and can support a wider variety and greater number of peripheral devices. Text-editing capabilities are implemented through software. Word processing service bureaus, which have developed text-editing programs for use on large-scale computers or larger minicomputers, represent a variant of the shared logic approach.

Figure 2-11. Shared processor, or shared logic, text-editing systems are special-purpose time-shared computers designed to support multiple input and output devices simultaneously. The Wang System 30, pictured here in a word processing center installation, can support up to 14 peripherals, features a hard disk with storage capacity equivalent to 4,000 typed pages, and will interface with optical character recognition or photocomposition equipment. (Courtesy: Wang Laboratories, Inc.)

Shared logic systems are especially attractive in applications where high capacity online storage is required and where several operators must edit different sections of the same document or assemble different documents from a centrally maintained file of previously entered text. At present, shared logic systems account for a relatively small percentage of installed text-editing systems, although significant growth is anticipated during the next several years. The high cost of such systems has necessarily proven a deterrent to acceptance in applications requiring only 1 or 2 workstations, but at $100,000 for a shared logic system capable of supporting 10 operators, the cost per workstation compares very favorably with that of stand-alone display-oriented systems. In addition, shared logic systems, with their more powerful information processing capabilities, are well suited to the implementation of integrated systems which combine text-editing with data processing and/or electronic mail. Such integrated systems, as discussed later in this book, are now attracting considerable attention and are expected to continue to do so over the next several years.

A frequently heard criticism of shared processor systems is that all workstations become inoperative when the central processor fails or otherwise malfunctions. A special variant of the shared processor system, called the *shared resource* system, overcomes this limitation. It combines stand-alone, display-oriented input stations, which are capable of independent functioning through locally available firmware or software and magnetic storage on floppy disks or other media, with other components, such as printers or large-capacity disk storage, which are shared by the various workstations. Such a configuration allows the system designer to obtain maximum utilization of the shared components—which in a stand-alone system remain idle much of the time—while obtaining some protection against processor failure. With their microprocessor-controlled autonomous workstations, shared resource systems are patterned after the distributed data processing systems described in Chapter Three.

Recording/Storage Media

As computer-like devices, all automated text-editing systems have an internal memory, manufactured from semiconductor materials, which provides temporary storage for text under immediate manipulation. The size of this internal memory varies. Most typewriter-oriented systems, for example, have a relatively limited memory that is perhaps sufficient to contain one or two lines of text (several hundred characters). The memory of the typical display-oriented system, however, will hold the equivalent of several pages, while shared logic systems may have internal memories sufficient to contain a multi-page report.

Regardless of size, the necessarily limited capacity of a given system's internal memory must be supplemented by auxiliary storage media and devices. As already noted, the earliest automatic typewriters used punched paper tape or belts to capture keystrokes for later repetitive playback. The MT/ST, with its magnetic tape cartridge, represented the first step toward today's virtually complete displacement of paper tape by magnetic media in text-editing applications requiring error correction and revision capabilities. The MT/ST cartridge was itself displaced by magnetic cards and tape cassettes, both of which have increasingly given way to flexible diskettes— the floppy disks which serve as the recording/storage medium in both text-editing and microcomputer systems.

As the name implies, magnetic cards are specially coated, tabulating size recording/storage media. As originally developed for use in IBM Magnetic Card Typewriters, each card has a capacity of 50 lines of up to 100 characters per line, for a total card capacity of 5,000 characters—the approximate equivalent of 2 double-spaced typewritten pages. During actual recording, however, a certain number of character spaces on each line are reserved for later insertions and other text changes. As magnetic card systems gained popularity, Redactron, Xerox, and others introduced cards with storage capacities approaching 10,000 characters.

Magnetic cards establish a one-to-one correspondence between recording/storage medium and the typed document, provided that the document is a few pages or less in length. Text-editing cassettes, by way of contrast, consist of 200 to 300 feet of computer-grade magnetic tape in a protective housing. While the standard cassette format is identical with that employed in dictation systems and other audio recording equipment, the tape itself is designed specifically for the recording of digital data rather than voice. Depending on the system, each cassette may record from 30,000 to 130,000 characters, the equivalent of as many as 25 magnetic cards. As is true of magnetic cards, some of the available character spaces are reserved for later insertions, so actual cassette capacity is invariably somewhat lower than the maximum. With most systems, however, the operator exercises some control over the amount of reserved space. In any case, a tape cassette offers much greater storage capacity than an equivalently priced number of magnetic cards. Cassettes are also better suited to the recording of reports and similar multi-page documents.

Reel-type computer tapes, which are similar to text-editing cassettes, limit the operator to serial access to particular portions of recorded text. In a retrieval and playback operation, movement from one end of a cassette to the other may take several minutes. Magnetic cards permit a kind of random access that is achieved only at the expense of storage capacity and with the aid of manual filing. Disk storage solved the problem of random access in computer systems and, while some shared processor text-editing systems employ conventional hard surface disks, the most advanced stand-alone systems rely on flexible or floppy disks for relatively inexpensive random access storage. A floppy disk is a magnetic-coated, circular-shaped piece of polyester on which data are recorded in a series of concentric tracks. The standard floppy disk measures 8 inches in diameter and will store 128,000 to over one million characters, depending on the recording method utilized. The so-called *mini-floppy* disk measures 5.25 inches in diameter and has approximately half the storage capacity of the standard version. Continuing improvements in coatings and manufacturing methods are expected to eventually increase disk storage capacities to more than several million characters. Both types of floppy disks are stored in a paper envelope to facilitate handling and to minimize the potential for damage to recorded data. When not in use, the disks may be maintained in storage cabinets or racks. For reading or recording, a disk is inserted into a specially designed drive which is typically built into the text-editor's control unit or input

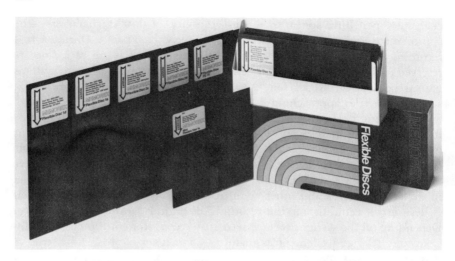

Figure 2-12. While older text-editing systems used magnetic cards or tape cassettes, most newer stand-alone systems utilize floppy disks in either the standard or mini-floppy version. The disks are stored in a protective paper envelope designed for direct insertion into specially-designed disk drives. (Courtesy: Memorex Corporation)

station. A special mechanism called a *read/write head* fits into an opening in the envelope, allowing access to any location on the disk surface in a few seconds or less. Given these relatively fast access times, floppy disks have become the recording/storage medium of choice in high performance text-editing systems. While magnetic cards and tape cassettes are still used in the majority of installed equipment, vendors previously committed to such systems are increasingly utilizing floppy disks in their newer products. Some floppy disk-based systems can be equipped with an auxiliary magnetic card or tape reader to facilitate the conversion of previously recorded text as older systems are replaced.

As briefly noted above, shared logic text-editing systems typically utilize hard-surface disks to obtain the multi-million character storage capacity required in conventional operations. In some cases, these systems employ conventional platter-type disks stacked in multiple units on a common spindle to form the disk pack configuration familiar to computer system users. Alternatively, a single-platter Winchester-type disk may be used. Winchester-type disk storage, which is often used with mini- and microcomputers, requires a mechanically less complex drive and is, consequently, less expensive than conventional disk systems. With some shared resource systems, as already discussed, individual workstations include a floppy disk drive for the local storage and processing of text that is not required by other operators.

Basic Text-Editing Capabilities

Regardless of the equipment configuration or recording/storage medium utilized, the operator typically communicates with a text-editing system through specially labelled keys mounted on either the input device or on the control unit. Certain keys initiate such basic system operations as recording, printing, and revising previously recorded text. Others allow the operator to search for, or advance to, a previously recorded word, line, or paragraph; specify document formatting instructions; and initiate any of the available special editing functions discussed later in this section. In display-oriented systems, a portion of the CRT screen or single-line display usually is reserved for prompting or other messages from the control unit. Such operator/machine interaction is not provided with most typewriter-oriented text-editing systems.

In terms of basic capabilities, all available text-editing systems allow the operator to print repetitive text automatically, correct typographical or other errors, and make insertions, without extensive retyping, in previously typed material. The automatic typing of repetitive text, as already briefly noted, can be accomplished by the most primitive systems and does not involve actual editing in the sense of text change or manipulation. Once keystrokes are recorded, the operator need only activate special command keys to replay the previously recorded material a sufficient number of times to produce the desired quantity of typewritten originals. With some systems, special instruction codes can be used to reprint text automatically on continuous paper stock without an operator in attendance.

Error correction and text revision capabilities distinguish text-editing from mere automatic typing. Magnetic media and internal semiconductor memories are easily correctable, since recording over a given portion of medium or memory erases its previous content. During input, typed characters are recorded in internal memory. Depending on the amount of available memory, these characters will be recorded on magnetic media at the end of one or several lines or pages. Errors can be corrected immediately by backspacing to the point of error and striking over the mistake with the correct character or characters, thus replacing the incorrect text on the medium or in memory. In display-oriented systems, the strikeover action removes the error from the screen as well. In typewriter-based systems, the strikeover results in a blemished first draft, but this does not matter since the operator's goal during the draft stage is to record all text correctly on magnetic media for later playback as a perfect typewritten original. To avoid blemished drafts, some typewriter-based systems employ the IBM Correcting Selectric with its dual-ribbon system which removes typographical errors from the paper and corrects them in memory or on the medium.

To correct errors detected during proofreading, or to make revisions in previously recorded text, the operator employs special keys and search commands to advance through memory or the recording medium to the point of error or revision. While unwanted text can be deleted routinely, the length of possible insertions varies from system to system. With the simplest magnetic card and cassette systems, as already noted, a predetermined number of character spaces are reserved during recording in anticipation of later insertions. Characters in excess of the reserved number cannot be inserted. Early equipment enhancements involved the addition of a second card gate or cassette drive to accommodate lengthier insertions by copying from one card or cassette to another. Newer text-editing systems, however, feature larger internal memories that make lengthy insertions much easier to perform. Text stored on a floppy disk is read into memory where the required insertions are made—the control unit manipulating the text is necessary to accommodate the additional material. The revised text then is recorded back onto a floppy disk. The previous version can be retained or deleted as required.

Advanced Editing Features

In addition to the basic ability to reprint and revise previously recorded text, many text-editing systems incorporate special features designed to facilitate input or enhance the appearance of printed output. The automatic centering of titles and alignment of decimal points in tabular presentations are examples of capabilities that simplify commonly encountered but time-consuming typing routines. Similarly, an increasing number of systems provide a calculator-style numeric keypad for rapid entry of quantitative data and, in the preparation of tabular presentations, permit the insertion or rearrangement of rows and columns. Several systems provide a dedicated key for paragraphing, thus replacing the combination of a carriage return and tab with a single keystroke. An increasing number of systems provide automatic numbering of pages and footnotes and will adjust those numbers, as required, to accommodate the rearrangement of paragraphs or other document restructuring.

To facilitate the location of text to be edited, most newer systems establish a directory that lists and briefly describes the documents recorded on a given floppy disk or other medium. Depending on the system, this directory may indicate the document title; the date it was created, printed, or revised, and by whom; its length and estimated printing time; and special security restrictions, if any. Once its location on a given disk is determined and the document is read into the processor's memory, the text can typically be searched for the first occurrence of a given character, word, or phrase. Several systems allow the operator to make *global* changes, correcting all occurrences of a given error. A few systems offer very special error correction capabilities, such as a dedicated key which transposes two letters in a single stroke, thus facilitating the correction of one of the most common typing errors.

Errors aside, text typed correctly once must often be retyped to accommodate format changes or to rearrange the order in which various text elements are printed. To simplify the creation of contracts, form letters, and other nearly identical typewritten originals, most systems provide some *document assembly* capabilities. While the specific approach to document assembly varies from system to system, pre-recorded text segments—such as the standard sections of a contract or proposal, or the individual paragraphs of a letter to be used in a mass mailing—are extracted automatically from a floppy disk or other storage medium and are combined for printing in an order specified by the operator. Alternatively, specified paragraphs or other text segments can be merged at printing time with pre-recorded variable data, such as the names and addresses of clients or the intended recipients of a mass mailing. Many new systems will search lists of names, selecting only those which meet certain operator-specified criteria. Thus, a mass mailing could be directed to persons residing in specified geographic locations. This *list management* capability is one example of a data processing function performed by text-editing equipment.

To change the appearance of a document without retyping, most text-editing systems permit changes in margins, line spacing, and line lengths. With some systems, text can be printed in double columns on a page. To enhance the appearance of printed output, many systems allow the operator to justify or otherwise control the contour of the right margin. Because such margin alignments and adjustments typically require hyphenation, a small but increasing number of systems feature automatic hyphenation capability based on either pre-programmed rules or an internally stored dictionary. A few systems are taking a similar approach to the automatic detection and correction of the most commonly encountered spelling errors.

TEXT-EDITING AND RELATED TECHNOLOGIES

Enhanced Typewriters

In most applications, the introduction of automated text-editing is accompanied by a complete or partial reorganization of clerical responsibilities. Secretarial employees—who previously may have performed a variety of office-related tasks in addition to typing—are divided typically into two groups: *word processing operators,* who concentrate on the production of typewritten documents free of the interruptions associated with general office work; and *administrative secretaries,* who answer telephones, make travel arrangements, maintain files, and perform other support work. Freed of all or most typing responsibility, the administrative secretary who previously served only one principal can now perform support work for several. In those offices where a secretary previously supported several principals, that number may be increased.

Given the prestige associated with having a dedicated secretary, many upper-level managers have strongly resisted the introduction of text-editing systems, preferring instead to retain the one-principal-to-one-secretary work relationship. Even where work responsibilities are reorganized in the manner described above, administrative secretaries may retain certain typing responsibilities, especially for memoranda, correspondence, or other relatively short documents that do not require extensive revisions. To meet the requirements of such applications, an increasing number of equipment manufacturers are developing typewriters designed to improve typing efficiency without a restructuring of clerical support or where administrative secretaries will perform occasional typing to supplement the capabilities of a word processing center. The earliest of these devices, the *memory typewriter*, features a captive magnetic recording medium capable of storing 50 to 100 typewritten pages for subsequent repetitive typing or moderate editing. Resembling the Selectric Typewriter in size and appearance, it simplifies document creation by allowing the operator to type at rough draft speed, confident in the knowledge that errors can be corrected easily prior to playback of a final copy from the magnetic recording medium. Priced at about $5,000 at the time this chapter was written, the memory typewriter was less expensive than more powerful text-editing systems, but too costly for the wholesale replacement of conventional office typewriters.

Figure 2-13. Important developments in office typewriters are reflected in four models from the IBM product line. The Electromatic (left), the first successful electric typewriter, was introduced in 1935. The Executive (background) was developed in 1941 as the first typewriter with proportional spacing. The Selectric (right), the first single element typewriter, was introduced in 1973. The Electronic Typewriter (foreground) is one of the three "intelligent" microprocessor-controlled models. (Courtesy: IBM Office Products Division.)

The *intelligent* typewriter, a product of more modest capabilities and price, is designed to simplify many of the most time-consuming typing routines. It can, for example, perform automatic centering and decimal alignment, provide a variety of horizontal character spacings, and store frequently used phrases or, in some cases, an entire page or two in a small internal memory. The intelligent typewriter lacks the power of a full text-editing system or even a memory typewriter but, at approximately $1,600 at the time this chapter was written, it costs only about $700 more than a conventional office typewriter.

Text-Editing and OCR

Actual editing—defined as the alteration or other manipulation of machine-readable text—is just one of three tasks text-editing equipment is designed to perform. The other two—input and output—may be performed by devices employing related technologies. Some information systems analysts, for example, contend that input—the task of converting text to the machine-readable form required for subsequent editing, storage, and printing—is an activity that does not warrant the use of a relatively expensive text editor. The initial keystroking of text, they maintain, can be accomplished more economically on a less sophisticated device. Some manufacturers permit greater flexibility in application design through a product line capable of being upgraded with compatible equipment of increasing power and price. Recently, however, the interfacing of text-editing and optical character recognition (OCR) equipment has permitted the preparation of text on a Selectric or other ordinary office typewriter equipped with an appropriate type font.[8]

While relatively new to word processing, optical character recognition is an established input methodology in data processing applications. As a computer input device, an OCR reader is essentially a scanner which converts human-readable typewritten characters to their machine-readable equivalents by analyzing and identifying light patterns reflected from typed documents. These characters, in coded form, are then transferred to, and recorded on, a magnetic tape or disk for later computer processing. In text-editing applications, secretaries, working from word originators' dictated recordings or longhand drafts, generate rough drafts of typewritten documents on their own typewriters at the fastest possible speed. These typed drafts—reviewed and marked for corrections or revisions—then are taken to an OCR reader operating online to one or more text-editing systems. The typed copy is scanned and the text is recorded on floppy disks or other media for later recall, editing, and printing. While older OCR readers required a special type font optimized for machine recognition, several newer readers can recognize elite, courier, and other popular typewriter fonts, although the typist often must observe special document formatting procedures.

While an OCR reader suitable for text-editing applications may cost

Figure 2-14. OCR page readers, like the Burroughs 1200 Series, enable office typewriters to prepare input for editing on word processing systems. The 1200 Series can interface with the most widely used stand-alone and shared logic text-editing systems and can accept documents prepared in several different type fonts. (Courtesy: Burroughs Corporation)

$25,000 or more, this approach can prove effective in reducing the number of required text-editing systems through the dispersal of initial typing. Traditional work relationships, with typing support located in close proximity to word originators, can be maintained, thus minimizing the resistance to change which is sometimes encountered in fully centralized text-editing installations. An OCR interface further contributes to the compatibility of text-editing systems of different manufacturers since text can be printed by one system and input, via an OCR reader, to another, otherwise incompatible system.

Text-Editing and Photocomposition

A preceding section of this chapter noted that the typical text-editing system includes a typewriter-quality printer capable, in most cases, of mono- or proportional spacing. There has been considerable recent interest, however, in combining text-editing with higher quality output in applications involving the production of reports, proposals, and similar documents. In addition to its more distinctive and attractive appearance, typeset text occupies only about 60 percent of the space required by its typewritten counterpart. The resulting compaction can reduce duplicating and dissemination costs significantly.

Most typesetting is now performed on photocomposers or other devices that generate individual characters photographically or electronically rather than by direct impression or by the setting of precast metal type slugs. The operating characteristics of such equipment are described in a later chapter of this book. For purposes of this discussion, however, it is important to note that typesetting equipment requires that text be encoded, together with format instructions, in machine-readable form. Meaningful savings in input preparation costs often can be attained if text which is initially recorded on word processing media is converted to a form suitable for typesetting, thus eliminating redundant keystroking. While the codes and formats required by text-editing systems and typesetters are typically very different, an interface between the two types of equipment can be accomplished in several ways. Some systems can be equipped, for example, with optional electronic connectors which translate formats and codes generated by a text-editing system into the format and codes required by specified typesetting equipment. Such connectors are available from manufacturers of both text-editing systems and typesetters, as well as from independent sources. Alternatively, text editing and typesetting capabilities can be combined in a single device which appears to the operator to be essentially a display-oriented text-editing system in which the typewriter-like output device is replaced by a photocomposing unit. When the text is entered, with format instructions, it is edited on a CRT screen prior to typesetting. The text itself can be recorded and stored on floppy disks or other magnetic media for later revisions.

Word and Data Processing

Given the obvious similarities between the design of text-editing systems and that of computer equipment, it should cause little surprise to find text-editing systems interfacing with computers and even acquiring computer-like capabilities. Most text-editing systems can be optionally equipped with telecommunications capability. Besides its utility in the electronic message system applications discussed in a later chapter, this telecommunications capability enables the text-editing system to emulate a computer terminal; these operating characteristics are discussed in the next chapter. This terminal-emulation feature can be especially useful in two types of applications:
1. In large-scale information storage and retrieval applications where text-editors will be used to prepare text locally for input to centralized disk or tape files; and
2. Where the text-editing system will be used to access remote data bases for text to be recorded, modified, and printed locally.

When functioning as a terminal, a text-editing system of course can be used to enter or activate computer programs for remote execution. Text-editing systems can, however, be viewed as special-purpose computer systems in their own right, and there is an increased tendency for text-editors to acquire local arithmetic, sorting, or other conventional computing capabilities.[9] In software-based stand-alone systems, most of which are microprocessor-based, manufacturers can add such capabilities without hardware modifications. Some newer systems are delivered with one or more blank keys in anticipation of such future software enhancements. It should be noted, however, that even with arithmetic and sorting capabilities provided as a standard or optional feature, most stand-alone text-editing systems cannot be programmed in the conventional data processing sense. Their computing capabilities are limited essentially to such operations as the automatic totalling of columns of numbers and the sorting of mailing lists into alphabetic or zip code order. This limitation is not necessarily characteristic of shared processor systems, some of which incorporate a fully programmable minicomputer. Several shared processor systems can be acquired with pre-written software suitable for such data processing applications as accounts receivable and payroll. When configured in this manner, these text-editing systems resemble the small business computer systems discussed in the next chapter. As noted in that chapter, many small business computers are capable of text-editing, thus providing an alternative approach to the integration of word and data processing.

3
Computers and Office Automation

Online terminals • Computers and management information • Small business computers

As discussed in Chapter One, computers have been applied extensively to the automation of "back office" paperwork processing since the mid-1960s. Today, most medium to large-size companies, government agencies, and institutions have computerized the posting, sorting, or other formerly manual clerical routines associated with such relatively "structured" operations as general ledger maintenance, accounts payable, accounts receivable, customer transaction processing, and inventory management. In these comparatively straightforward data processing applications, computers have been the source of significant improvements in speed and efficiency.

Computers, however, have proven somewhat less successful in the relatively "unstructured" applications encountered in the "front office." The preceding chapter noted, for example, that computer-based text-editing systems generally have failed to gain widespread office acceptance, largely because they are designed more often for use by programmers than for secretarial personnel. Through the mid-1970s, computer equipment manufacturers and data processing personnel took little interest in such office applications. University-based computer science departments, business administration programs, engineering schools, and other sources of training for information systems professionals rarely incorporated office automation courses in their curricula. In the last several years, however, the data processing community has become increasingly aware of the importance of well-designed office information systems. This awareness is reflected in the extensive coverage now routinely given both to office automation in data processing periodicals and in the growing involvement of computing center directors and other data processing personnel in the planning and implementation of automated systems. The computer industry now recognizes the office as a large, attractive market with unique requirements and recently has begun developing products and services especially for it. This chapter reviews those developments, emphasizing the hardware and software which offices utilize to obtain online access to remote computers and the increased presence of computing equipment in the office itself.

ONLINE TERMINALS

Background

When the "back office" operations were first computerized in the 1960s, the computer itself often was located in a centralized facility outside of the office, but was physically accessible to it. Office personnel prepared and packaged source data for submission to a centralized keypunch department or, alternatively, keypunched or otherwise converted the data to machine-readable form. The data, on punched cards or magnetic tape, then were batched for processing by the computing center at regularly scheduled intervals, and the results were returned to the office in the form of paper printouts. While such batch processing systems, as previously noted, eliminated many manual clerical operations, they often contributed to information storage and retrieval problems. Computer-generated paper reports, for example, often prove difficult to handle and maintain. Computer-output-microfilm (COM) technology, discussed in the next chapter, developed, in large part, as a remedy for this problem. Similarly, turnaround times for batch processing and report generation sometimes prove so long that the resulting information loses its value as a support for decision-making or other office-related activities. Lateness aside, batch-processed reports often provide the prospective user with too much information, making it difficult to accommodate specific information requirements. Finally, batch processing effectively limits the user to computer resources located within a reasonable geographical distance of the office.

Although a large number of office applications still employ batch processing, the development of time-sharing computer operating systems and their widespread application during the 1970s successfully addressed many of these problems. Voluminous paper reports can be replaced, for example, by a combination of screen displays and selective printouts. The turnaround time for a given information-processing task can be reduced to seconds, thus providing information in a time frame appropriate to ongoing decision making. The development of conversational, interactive programs allows the computer user to request specific items of information rather than an entire report. Prior limitations of geographic proximity to computing resources are likewise removed in most cases. The availability of time sharing, combined with improvements in data base and telecommunications technology described later in this chapter, have increased significantly the range of information-processing resources available to support office operations.

The online terminal is the critical piece of office equipment in time-sharing computer applications. Because terminals play such an important role in the various aspects of office automation discussed throughout this book, knowledge of their technical and operating characteristics is essential to the design of effective office information systems. The next sections of this chapter describe and discuss the most important features and functions of available online terminal equipment.

Terminal Configurations

For purposes of this chapter, an online terminal is defined as a device which permits the interactive transmission of data to, or the reception of data from, a computer or other information-processing machine via electronic digital pulses transmitted over connecting wires. As such, terminals are an important component in online computer systems and in various other automated office products, including the text-editing systems discussed in the preceding chapter, as well as the micrographics retrieval systems and electronic mail and message systems described later in this book. As is true of computers, text-editors, and other information-processing equipment, an online terminal can be viewed as an integrated configuration of up to four groups of components: (1) a central processor or control unit; (2) internal memory and, in some cases, auxiliary storage; (3) an input mechanism; and (4) an output mechanism. Newer terminals invariably utilize microprocessors as controllers and are categorized, in part, by their inherent information-processing capabilities. The simplest terminals have relatively primitive microprocessors and are nonpejoratively described as "dumb" to reflect the fact that they must derive their information-processing capabilities from an external computer. "Intelligent" terminals, by way of contrast, feature a more powerful microprocessor and, being user-programmable, have information-processing capabilities independent of external computers. Intelligent terminals are actually a variant form of microcomputer and, as such, will be discussed later in this chapter. A third category of terminal equipment, loosely described as "smart," is not user-programmable in the conventional data processing sense, but incorporates microprocessor-based features not found on a simple "dumb" terminal. These features include a wide range of display attributes, transmission speeds, or printing capabilities; an internal memory for simplified data preparation or temporary storage of received data; and the ability to redefine the functions of specified keys to meet special application requirements. Most newer online terminals fall within this group and, at the time this chapter was written, were priced about 25 percent higher than their "dumb" counterparts.

In terms of their output mechanisms, online terminals are commonly divided into display and printing devices. The latter group, often called *teleprinters* or *teletypewriters*, are typewriter-like machines which provide the operator with paper output. The most commonly encountered display terminal features a cathode-ray-tube (CRT) screen, although a few display terminals incorporate more compact plasma screens. The number and popularity of such flat panel displays are expected to increase over the next decade. Display terminals are described collectively as "soft copy" devices, although they can be connected to an auxiliary printer where paper output is desired.

Regardless of their output mechanisms, online terminals can be purchased in any of three basic models:

1. The most common, the *Keyboard Send-Receive* (KSR) configuration, can transmit data to, and receive data from, a computer or other information-processing machine. Transmission is initiated through a keyboard. The output mechanism is a printer or screen. The keyboard, the distinguishing feature of the KSR configuration, may be either integral to the terminal chassis or detached. In the latter case, the keyboard is connected to the

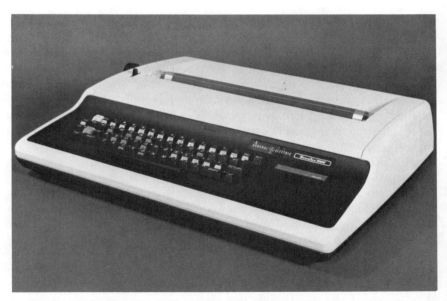

Figure 3-1. Keyboard Send-Receive (KSR) terminals can both transmit and receive data. The GE TermiNet® teleprinter, pictured above, is a versatile, microprocessor-controlled device well suited to desk-top operation in an office environment. (Courtesy: General Electric Company.)

Figure 3-2. The most commonly-encountered display terminal is a Keyboard-Send-Receive device with a cathode ray tube (CRT) screen. In many applications, the display terminal is configured with a Receive-Only printer for selective hard-copy output. (Courtesy: Lear Siegler, Inc.)

terminal by a cord and can be positioned to suit the operator's taste. Several keyboard layouts are available, the most popular resembling that of a Selectric Typewriter with additional keys that initiate and control data transmission. In some cases, a calculator-style numeric pad is provided for the simplified entry of quantitative data.

    2. A *Receive-Only* (RO) terminal is capable of displaying or printing data received from a computer or other information-processing machine, but cannot itself transmit data. Physically, an RO terminal is distinguished by the absence of a keyboard. Receive-only teleprinters are often attached to KSR display terminals in applications requiring selective paper output. They are likewise widely used as printers in display-oriented text-editing systems.

    3. The *Automatic-Send-Receive* (ASR) configuration consists of a KSR terminal with peripheral attachments or internal memory circuits which permit the offline preparation of data prior to transmission or the capture and storage of incoming data. Historically, ASR terminals have incorporated a paper tape reader/punch. A growing number of ASR terminals, however, feature a magnetic tape cassette recorder or a floppy disk drive, and several newer portable terminals feature internal bubble memory storage. In this hardware configuration, ASR terminals resemble the stand-alone text-editing systems described in the preceding chapter and the microcomputer systems

Figure 3-3. While CRTs now dominate the display terminal market, gas plasma and other flat panel displays are expected to gain popularity during the 1980s. Such flat panel terminals are much more compact than CRT devices. The General Digital VuePoint terminal, for example, occupies relatively little desk space and features a touch sensitive screen as an alternative to conventional keyboard input. (Courtesy: General Digital Corporation)

described later in this chapter. They lack, however, the firmware or software required for word or data processing.

Communication Characteristics

As a communications device, an online terminal is designed to transmit and/or receive digital data in the form of an electrical signal which consists of a series of on/off pulses. Together a pre-determined number of these pulses—which, in data processing terminology, are called binary digits or *bits*—form individual characters intended for transmission and/or printing. In the most prevalent mode of terminal-to-computer communication, called *serial asynchronous* transmission, the individual bits that make up a given character are transmitted in sequence, framed (preceded and followed) by additional bits that separate successively transmitted characters from one another. These additional bits are required because a given terminal is typically not synchronized with the computer or other device with which it is communicating. *Synchronous* terminals, which do not require framing bits, are capable of faster transmission, but are more complex and expensive than their asynchronous counterparts.[1]

Several different code systems define the number and position of the

Figure 3-4. Integrated circuits and microprocessor technology are increasingly blurring the distinctions between information processing systems. The NEC ASTRA 205 is technically a microcomputer. When equipped with communications capability, however, it can function as an "intelligent" ASR terminal. When furnished with appropriate software, it becomes a word processing system. (Courtesy: NEC Information Systems.)

bits which online terminals transmit to represent individual characters. In the United States, the most prevalent code system is the *American Standard Code for Information Interchange* (ASCII). The full ASCII (pronounced "askey") code set includes 95 printable characters, including the upper and lower case alphabet, the digits 0 through 9, 32 punctuation marks and special symbols, and the blank space. The remaining 33 characters represent control codes that initiate some action in the terminal or the computer with which it is communicating. Some terminals, in the manner of the once-dominant Teletype Model 33, transmit a subset of 96 ASCII characters, omitting the lower case alphabet and certain special symbols. For Teletype-compatibility, many full-ASCII terminals permit the operator to restrict transmission and reception to this subset. In addition to the ASCII code, some teleprinters can be configured to transmit the APL/ASCII code with its special characters designed to support the APL programming language. Terminals designed for use with IBM computers may transmit in any of the three other code configurations: Correspondence, the code used by IBM Selectric Typewriters; the Extended Binary Coded Decimal Interchange Code (EBCDIC); and the Binary Coded Decimal (BCD) code. For international use, some terminals transmit the Baudot code, a slower configuration that is also used in telex communications.

In manufacturers' specification literature, terminal speed is reported typically in either of two measures: *bauds* or *bits per second*. The interchangeable use of these measures—a common practice—is the source of some confusion and potential inaccuracy. Terminals communicate data by altering the amplitude, phase, or other condition of a telephone line or other transmission facility. Successive alterations in line conditions represent the on/off pulses (bits) used to encode individual characters. The term *baud* denotes the number of times a given line condition changes in a second. *Bits per second*, however, is a measure of the rate at which successive individual bits are transmitted or received by a given terminal. As long as only one bit is reflected in each line condition change, a terminal's baud and bit rates will be equal. This is true of most online terminals, but certain complex, high-speed devices are capable of transmitting more than one bit per baud.

As indicated in the preceding discussion of asynchronous transmission, the bits used to represent individual characters—sometimes called *intelligence bits*—are preceded and followed typically by *start* and *stop* bits which compensate for the lack of synchronization between transmitting and receiving devices. In addition, most ASCII terminals transmit an additional *parity* bit with each character. This parity bit is used to monitor transmission errors. The total number of intelligence and other bits required to transmit individual characters varies with the coding system used and the terminal's speed. Most terminals are designed to operate over ordinary telephone lines at one or more speeds in the 75 to 1200 bits per second range—110, 300, and, increasingly, 1200 bits per second being the most widely used. With leased telephone lines, transmission speeds ranging from 2400 to 9600 bits per second are possible.

There is a simple formula that can be used to relate the number of bits transmitted per second to the number of characters transmitted per second. With asynchronous ASCII terminals communicating at 110 bits per second, 11 bits are used to represent each character; at faster speeds, 10 bits are used. Thus, the 3 most popular transmisson speeds of 110, 300, and 1200 bits per second are the equivalent of 10, 30, and 120 characters per second.

Regardless of operating speed, most online terminals can transmit in either or both of two line modes. In the *half-duplex* mode, data entered at the terminal keyboard are printed locally and transmitted to a remote computer or other receiver. Reception and transmission cannot, however, occur simultaneously, and it is not possible to interrupt the reception and printing of data once they have begun. In the *full-duplex* mode, data entered at the terminal keyboard are transmitted to a remote computer which echoes them back for printing and verification of accurate reception. The full-duplex mode permits simultaneous transmission and reception. The decision to use the half- or full-duplex mode is dictated by the computer system with which the terminal must communicate.

Line Interfaces

As discussed in the preceding section, the output of an online terminal is an electrical signal consisting of a series of discrete digital pulses. When the terminal is located in close proximity to a computer or other receiver, this digital signal can be transmitted over customer-owned wires. In such applications, the successive transmitted bits are represented by the presence or absence of electrical current. Most office applications, however, require online communication with a computer located at some remote point and use telephone lines as the communication medium.

Unfortunately, most telephone circuits were designed for the transmission of the continuously varying analog signal characteristic of the human voice. The terminal's digital signal consequently must be converted to analog form prior to transmission. This conversion, which is called *modulation,* is performed by a *modem,* an electronic device which serves as an interface between a terminal, a telephone instrument, and a telephone line. An alternative form of modem, called an *acoustic coupler,* establishes a temporary connection with the telephone network through the telephone handset itself. While a conventional modem establishes a direct electrical connection with the telephone network, an acoustic coupler converts the terminal's electrical signal to audible tones for transmission through the telephone's speaker. Incoming signals likewise are directed to the telephone's receiver. Operation of the two interface devices is otherwise similar. Using a telephone, the operator dials the number of the desired computer, then activates an appropriate modem switch or inserts the telephone handset into the acoustic coupler. With many computer systems, the terminal is connected continuously to the computer, and dialing the telephone is unnecessary.

While some terminals can be ordered with an integral modem or acoustic coupler, most cannot. In such cases, the teleprinter manufacturer provides a receptacle and/or power cord designed to interface with designated Bell System Datasets or equivalent competitive modems or acoustic couplers that the terminal user must rent or purchase separately. The connection of terminals to modems and acoustic couplers is simplified greatly by widespread adherence to the Electronics Industry Association RS-232C interface or the recently developed RS-449 interface specification. When a non-Bell modem is used, an additional interface called a Data Access Arrangement (DAA) may be required prior to interconnection to the public telephone network. The DAA, which is supplied by the telephone company, is designed to protect the telephone facilities from potentially harmful power surges.

Figure 3-5. Because most telephone circuits were designed for the transmission of the continuously varying analog signal characteristic of the human voice, the digital signal generated by terminals, computers and related equipment must be converted to analog form prior to transmission. This conversion, called modulation, is performed by a modem, an electronic device which serves as an interface between a terminal, a telephone instrument and a telephone line. (Courtesy: Racal-Milgo, Inc.)

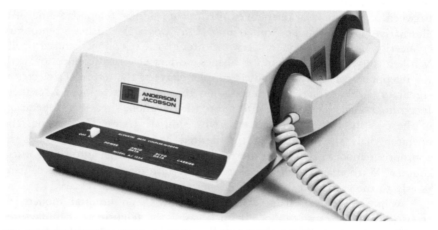

Figure 3-6. An acoustic coupler is a variant form of modem which establishes the line connection through a telephone handset. (Courtesy: Anderson Jacobson, Inc.)

Output Characteristics

An earlier section of this chapter noted the division of terminals into two broad groups according to their output characteristics—teleprinters, which provide paper printouts, and display terminals. As hardcopy output devices, teleprinters can be divided into two broad groups: *impact* printers and *non-impact* printers. Impact printers, as the name implies, employ a mechanism which drives an inked ribbon into paper—or paper into an inked ribbon—to print individual characters. Classifying impact printers by character generation, they can be subdivided into two broad groups: *dot matrix* devices and *embossed typeface* printers. Dot matrix devices feature a printhead which consists of a column of wires that are driven selectively into a sheet of paper. The characters themselves are formed from a pattern of discrete dots. Depending on the model, the matrix size ranges from five rows by seven columns to seven rows by nine columns. Even at its best, the resulting print quality, while acceptable for applications requiring only legibility, is definitely inferior to that obtainable with an IBM Selectric or an equivalent office typewriter. Quality may be especially marginal for the smaller characters in the lower case alphabet.

Yet, the typical dot matrix teleprinter is a versatile device that offers several significant advantages. It can print a wide range of different characters. On receipt of incoming data, the teleprinter's microprocessor-controller consults a table of character definitions stored in internal memory circuits to determine the order in which the print wires are to be activated. By merely modifying or replacing the memory circuits' contents, the device's

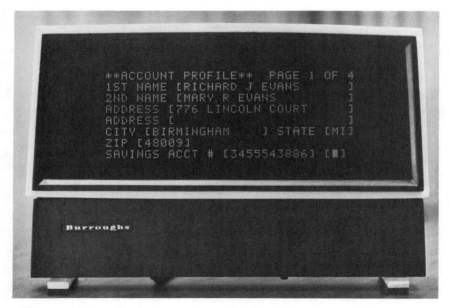

Figure 3-7. Most display terminals and many teleprinters utilize the dot matrix method of character formation in which individual characters are shaped out of selectively illuminated or printed dots. Output quality depends on the size of the dot matrix. (Courtesy: Burroughs Corporation.)

printable character set can be enhanced or altered. Thus, several dot matrix teleprinters can store definitions for two entire character sets—ASCII and APL, for example—or allow the user to generate customized characters. Equally important, microprocessor-controlled dot matrix devices are the fastest teleprinters currently available, being capable of sustained speeds in excess of 180 characters per second.

Although they offer versatility and high-speed operation, dot matrix devices cannot match the output quality of teleprinters, which generate characters from embossed typefaces. The best of such devices employ the daisywheel-shaped printing element employed in the word processing printers discussed in Chapter Two. As an equally effective alternative to the daisywheel, several teleprinters use interchangeable, rotating print cups. These typewriter-quality teleprinters operate at speeds in the range of 30 to 55 characters per second.

Nonimpact teleprinters employ a sheet of sensitized, coated paper which changes color in areas to which heat is applied. Characters are formed from a dot matrix by selectively warming a column of wires which are brought into contact with the paper. This thermal printing method has been applied very successfully to compact desk-top and portable teleprinters. Because hammers, rotating print elements, or other impact mechanisms are not used, thermal printing is quiet and reliable. A special, relatively expensive paper is required, however, and multiple copy production is impossible. Despite recent word processing applications of plain paper printers that use xerographic or ink-jet printing, such alternative non-impact technologies have yet to be utilized in teleprinters.

Regardless of the printing technology employed, the industry-prevalent horizontal spacing is 10 characters per inch (the equivalent of 10 pitch typewriter spacing), which provides for an 80-character line on an 8.5-inch-wide sheet of paper and a 132-character line with 14-inch-wide paper. Many teleprinters provide additional horizontal spacing options which can be selected by the operator at output time. Twelve characters per inch is, for example, the spacing used by typewriter-quality teleprinters equipped with an elite or courier printing element. Further compression of horizontal spacing to 16.5 characters per inch is possible with some teleprinters, thus permitting the printing of 132 characters on 8.5-inch-wide paper. The industry-prevalent vertical spacing is 6 lines per inch or 66 lines per 11 inches of paper. Paper requirements themselves depend on the design and width of the teleprinter's platen. Some teleprinters employ a friction-feed typewriter-style roller designed to accept paper in the form of single sheets or continuous rolls. More common, however, is the forms tractor, a mechanism designed to accept single or multi-ply fan-fold, edge-punched paper.

With regard to display terminals, the most commonly encountered output format consists of 24 lines of up to 80 characters each. This is smaller than the output format provided with the full-page display-oriented text-editing systems discussed in the preceding chapter. As is true of text-editors, however, a cursor—a special character in the shape of a rectangular block or underscore—marks the position where the next character will be entered. Successive lines of entered or received data begin in the leftmost position of the first row and are displayed until the screen is full, at which point the display typically scrolls upward to allow additional lines to be entered in the last row. Some display terminals are equipped with random-access memory (RAM) circuits which retain data which have scrolled off the screen. By activating specified controls, the operator can review previously displayed data. Depending on the particular model, these so-called "smart" terminals may provide memory capacity in the range of two to eight screens.

Most display terminals, as noted earlier in this chapter, are CRT devices which employ the dot matrix method of character generation. Upper and lower case character capability is normal on most newer equipment, and several terminals can be configured for the display of foreign characters or customer-specified special symbols. In terms of visual attributes, the display itself is typically of negative polarity, consisting of light characters on a dark background—a screen presentation described as *standard video*. Many CRT terminals allow the operator the option of displaying dark characters on a light background—a screen presentation known as *reverse video*. With most newer terminals, a filter in front of the screen reduces the noticeable glare which frequently contributed to complaints of operator eye fatigue in earlier CRT-based computer systems. Additional dials or other controls allow the operator to further adjust the contrast and intensity of the display to suit personal preference. The most sophisticated terminals allow the operator to selectively adjust the intensity and/or polarity of portions of the display, to highlight, for example, certain data items. With regard to some terminals, this highlighting effect can be enhanced further by causing specified data items to blink. These visual attributes can prove very effective when combined with simple line drawing capabilities in the presentation of tabular data, bar charts, histograms, and similar business-type graphics. Such terminals should not be confused with the more elaborate graphics displays described briefly in a later section of this chapter.

To simplify data entry applications, most "smart" terminals will display a formatted screen with labelled areas or fields into which the operator key-enters data, advancing from one field to the next using the tab key. Field labels and other selected portions of the screen can be programmed to blink in higher or lower intensity, or free from operator modification. This formatted display capability typically is utilized in conjunction with buffered or block transmission in which an entire screen or portion of a screen first is entered into the terminal's internal memory, then edited as required, and subsequently transmitted to a remote computer. To facilitate error correction and data modification prior to transmission, terminals capable of operating in this manner typically include dedicated function keys that permit the insertion or deletion of lines and/or characters. These editing capabilities are, however, quite primitive and should not be confused with the more sophisticated functions of true text-editing systems.

While they offer a wide range of capabilities and are increasingly designed with attention to human engineering considerations, the acceptance of display terminals is often limited by their inability to generate paper copies of data for future reference or dissemination. Most display terminals, however, will, support an auxiliary paper printer. As previously discussed, a Receive-Only (RO) teleprinter is often configured with a Keyboard-Send-Receive (KSR) display terminal to provide an essentially video-oriented device with selective hardcopy capability. The attachment of auxiliary printers is simplified by adherence to the Electronics Industry Association RS-232C standard interface described earlier.

COMPUTERS AND MANAGEMENT INFORMATION

Data Base Technology

In computer-based information systems, an online data base can be broadly defined as an integrated accumulation of machine-readable data maintained on one or more direct access storage devices.[2] The emphasis on integration distinguishes a data base from a data file. The latter term typically denotes an isolated accumulation of data designed to meet the specific requirements of one or several application programs. A data base, by way of contrast, is a common repository of data which many application programs can utilize. In the 1960s and early 1970s, as previously discussed, the creation of separate data files for each application program was standard data processing practice. While separate data files are still widely utilized, data base concepts developed in the mid-1970s as an outgrowth of many organizations' awareness of their substantial investment in, and continuing dependence on, their various accumulations of machine-readable data.

From the standpoint of productivity improvement, the idea that a manager or other white collar worker, equipped with only an online terminal, can obtain rapid, in-office access to a centralized collection of corporate or institutional data is obviously attractive. As a result, a strong data base orientation is an increasingly dominant characteristic of management information systems.[3] But, while online data storage and retrieval has long been recognized as a potentially important support for management decision making and related activities, the scope of online systems has been restricted necessarily by the limitations of available hardware and software. To conserve computer processing time and online storage resources, many management information systems continue to rely heavily on batch-processed paper reports as an adjunct to online access. Such reports are, for example, widely used for the recording of historical data that have been transferred from online disk storage to offline magnetic tape. Similarly, the speed and other operating characteristics of the central processor limit both the number of terminals that can be online at a given time, as well as the time required to respond to data retrieval requests. From the software standpoint, the development of online information retrieval programs requires considerable time and effort. In the past, such programs too often have been developed for a single application, operated on only a subset of an organization's available data, and offered a relatively narrow range of information-retrieval capabilities.

These limitations are in the process of being overcome through hardware and software advances that have important implications for office automation. From the hardware standpoint, computer technology is increasingly able to support large-scale online information systems. The cost/performance characteristics of newly introduced central processors have improved steadily by all available measures. Increased operating speeds and computational capabilities are reflected in the ability to support many terminals and in greatly improved response times. Equally important, the cost of online direct access data storage has dropped significantly and is likely to continue to fall. With regard to magnetic disks, the most prevalent form of direct-access storage, the cost per bit stored has decreased steadily since the early 1970s, making it possible to establish data bases consisting of hundreds of millions of characters and to maintain them online for longer periods of time. As a supplement to magnetic disks in very large applications, several available mass storage systems use magnetic data cells or other media to maintain billions of characters online. While the initial capital investment for such systems is substantial, the cost per bit stored is low.[4]

With the introduction of optical disks, additional substantial reductions in online storage costs are expected in the next several years. Such disks, which now exist in prototype, store data in combinations of encoded bits which are recorded and read by lasers.[5] They will offer an economical online alternative to magnetic tape and the existing mass storage systems described above. These optical disks differ from the image-oriented video disk systems described later in this book in that they will store character-coded, computer-processible data rather than pictures of documents. Unlike available magnetic disks, data recorded on optical disks will not be erasable. It is thus likely that magnetic and optical disks will coexist in online information systems, with optical disks serving as a long-term online repository for static data retired from magnetic disks.

Historically, the computer industry has been able successfully to overcome most of the hardware limitations which have impeded the development of increasingly sophisticated information-processing systems. Software problems, however, have proven more difficult to resolve.[6] From the programming standpoint, the retrieval of specified data from a very

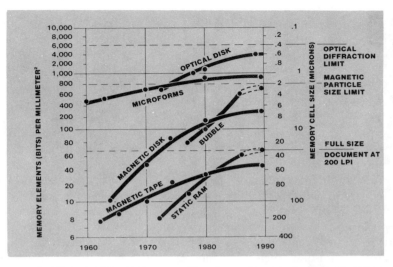

Figure 3-8. Historical limitations on the size and cost of computer memories are in the process of being overcome. With magnetic disks, currently the most prevalent form of direct-access storage, capacity has increased while costs have decreased steadily for the past decade. Additional substantial increases in online storage capacity, with associated cost reductions, are expected in the next several years with the introduction of optical disks which store data in combinations of encoded bits which are recorded and read by lasers. (Reprinted from the *Journal of Micrographics,* July/August 1980.)

large data base is rarely an easy task and, given the highly reactive and unstructured nature of much office work, it often proves very difficult or even impossible to fully anticipate a given manager's information-retrieval requirements. Conventional data processing approaches, involving the writing of customized application programs which produce printed reports or support limited online retrieval from a subset of available data, may prove adequate for the transaction-processing applications characteristic of back office operations but cannot satisfy the front office's *ad hoc* information requirements. The information retrieval needs of a customer service representative can, for example, typically be satisfied by periodic printed listings of order status information supplemented by online access to a small data file which reflects recent changes in the status of specific orders. A marketing manager or production planner, however, may need to quickly identify those customers who routinely purchase specified quantities of a given product or those products for which orders consistently exceed availability. Where such information retrieval requirements occur unpredictably or infrequently, they cannot be effectively and economically accommodated by custom-developed programs. Even when higher level programming languages such as COBOL,

FORTRAN, or PL/1 are used, the conventional software development process remains time consuming and relatively inflexible.

In some cases, however, a broad range of information retrieval requirements can be quickly and effectively handled through a special type of software called a *Data Base Management System* (DBMS)—a complex set of programs designed to facilitate the establishment and utilization of a data base. Data base management systems are available for implementation on the most commonly encountered large-scale computers, as well as on some of the smaller computers described later in this chapter or leased, in prewritten form, from computer equipment manufacturers or software development companies. The typical data base management system includes a series of programs which organize and load a user's data onto disks or other direct access storage devices in a manner compatible with later retrieval. For example, the system may: (1) establish and maintain various indexes to stored data; (2) eliminate the inconsistencies and redundancies characteristic of data maintained in separate files and; (3) provide a common data repository that various application programs can access. Most important for purposes of this discussion is the ability of most data base management systems to accommodate a wide range of *ad hoc* retrieval requests through specially developed *query languages*. Unlike conventional programming languages which require the specification of detailed instructions which the computer must follow to identify and retrieve data, query languages allow a user working at an online terminal to initiate information searches by key-entering a series of commands accompanied by specified retrieval parameters. While they do not permit the use of natural language in the expression of information-retrieval requirements, query language commands are typically easy to learn. The use of query languages does not require the training in algorithmic thought nor memorization of elaborate sets of rules that characterize many conventional programming languages. For example, a manager seeking employees suitable for a newly created position quickly and conveniently can initiate a search of a personnel data base for the names of individuals with specified educational backgrounds, a minimum number of years of service, work experience in particular activities, or various combinations of the above. The ability, through query languages, to initiate complex searches involving the logical coordination of several or many retrieval parameters is a particularly useful feature of data base management systems. It enables such systems to satisfy information retrieval requirements which either cannot be anticipated or occur too infrequently to justify customized programming.

Online Data Banks

The preceding section discussed online access to a machine-readable data base created and maintained by an organization for its own use. As used in this section, the term *data bank* denotes a collection of machine-readable data created by one organization for use by others. In most cases, the creator of the data bank also provides pre-written programs that permit the retrieval of specified items of information. These programs constitute a data base management system which includes a query language. The data itself usually is maintained on computers operated by the creating organization or a third party. Users pay a subscription fee and/or other charges to access the data bank from an online terminal.

While office workers rely most heavily on data that are created and maintained locally, there are times when external information resources must be consulted. As previously discussed in Chapter One, the performance of white collar workers generally is enhanced by information acquired from such sources as: (1) technical books and journals; (2) publications of professional and trade associations and; (3) newspapers and popular magazines. In addition to their general value, such published information sources are sometimes an integral component of white collar work. Scientists, for example, must have access both to patent information as well as to the latest technical publications in their areas of specialization. Attorneys must consult published collections of laws and court decisions routinely. Investment analysts must have access to the annual reports of corporations and related published documentation. The usefulness of such published information sources is not limited to research workers or specialists, but extends to practitioners in fields as diverse as engineering, public affairs, and medicine. Historically, however, the printed indexes and other sources which provide access to publicly available information contained in books, magazines, and newspapers have been maintained in an organization's library or technical information center rather than being conveniently and immediately available in the office. Many smaller organizations, and a surprising number of larger ones, do not provide their office workers with internal formal library support. In such instances, workers may maintain personal files of newspaper clippings and magazine articles or rely on public libraries or other external agencies. The result is time wasted in retrieving required information and a general tendency to avoid the utilization of publicly available information sources. In both cases, there is a potentially adverse impact on work performance and productivity.

Online data banks overcome these problems by making published information sources conveniently and rapidly available to any office equipped with a terminal.[7] A number of online data bank services are currently operational. They differ considerably in scope and range of available data. With few exceptions, they can be used effectively by office workers with modest training. As an example, the need to consult newspapers and popular magazines, as well as more technical publications, was mentioned previously as an important component of some office activities. Many organizations routinely scan such sources for information about topics of significance to their daily or long-term operations. Such topics may be monitored on a continuing basis for months or years, with a resulting significant expenditure of labor. Similarly, some organizations scan newspapers, magazines, and other published sources for any mention of their own activities. In the pharmaceutical industry, for example, it is common to monitor published information sources for references to an organization's products or their generic equivalents. As an alternative to the time-consuming, arduous procedure of manual scanning, the New York Times Information Bank permits subject or other retrieval of articles published in the *New York Times* and various other newspapers and magazines. With regard to articles from the *Times* itself, the user receives a list of citations to articles (including the title, author, and date of publication) plus a useful summary. Information about articles from other sources is limited to the citation. Similarly, the Dow Jones News/Retrieval Service provides online access to Associated Press and United Press International news dispatches. The entire text of the dispatch can be printed or displayed at the user's terminal. The NEXIS system, offered by Mead Data Central, is a unique data bank service that permits the retrieval of newspaper articles by a wide range of parameters. The data bank includes the entire text of the article, not merely a summary. As such, it entirely eliminates the need to access printed information sources. Similar services are either in development or being offered on a restricted basis by various newspaper publishers.[8] In most cases, these data banks are economically derived as a by-product of newsroom automation.

The New York Times Information Bank and the Mead NEXIS system are examples of data banks which are created and offered for sale by the same organization. MEDLINE, an online information service developed by the National Library of Medicine, is another example of a data bank made available by its creator. More often, however, the creator of a data bank makes it available through one of the services which specialize in the provision of online access to data banks created by others. Three such services are available worldwide: the DIALOG system, developed by Lockheed Retrieval Service; the ORBIT system, created by System Development Corporation; and the online search service established by Bibliographic Retrieval Services, which uses a modified version of the STAIRS information-retrieval software developed by IBM. These services all provide a single query language which can be utilized to access a wide range of data banks obtained from various sources.

The Lockheed DIALOG system, the most comprehensive of these services, will be used as an example of the types of data banks and search capabilities offered. Most of the available data are bibliographic in nature—that is, they consist of citations to books, journal articles, and technical reports. A search typically culminates in a list of citations which can be printed either at the user's terminal or offline for delivery by mail. The latter alternative is preferred if the information need is not urgent or if the citation list is lengthy. The system's retrieval capabilities are very powerful

```
File6:NTIS      64  81/Iss07
(Copt. NTIS)
        Set  Items  Description   (I=OR;*=AND;-=NOT)
? select electronic
        1  15487   ELECTRONIC
? select mail
        2    203   MAIL
? combine 1 and 2
        3     44   1 AND 2
? t 3/3/1
3/5/1
Electronic Message Systems:  The Technological, Market and Regulatory
Prospects
Kalba Bowen Associates, Inc., Cambridge, Mass. ** Massachusetts Inst. of
Tech., Cambridge.  Center for Policy Alternatives. *Federal Communications
Commission, Washington, D.C. Office of Plans and Policy.
AUTHOR: Kalba, Konrad K.; Sirbu, Marvin A. Jr; de Sola Pool, Ithiel;
E1971A4    Fld: 17B, 45C*    GRAI7817
Contract: FCC-0236
Monitor: FCC/OPP 78/0236
Prepared in cooperation with Massachusetts Inst. of Tech., Cambridge.
Center for Policy Alternatives
```

Figure 3-9. Online data banks, such as the Lockheed DIALOG retrieval service, support interactive information searches. In the example depicted here, the NTIS data base is searched for reports dealing with electronic mail. In the simplest approach, keywords are first selected then combined to create a set of candidate citations which can be printed at the terminal or offline.

and include the ability to perform complex searches by author, subjects, or other parameters. In addition, the text of citations can be searched for specific words or combinations of words. Some citations include an abstract, but not the full text of the article or report cited. In most cases, however, a copy of the actual document can be ordered online for delivery by mail. The total number of data banks accessible through the DIALOG system exceeds 100 and is continually increasing. Many of the available data banks are highly specialized in scope, being limited, for example, to a particular type of document, such as technical reports, or to a particular subject. A number of the available data banks, such as the Magazine Index or the Public Affairs Information Service index, are sufficiently general to be applicable to a broad range of information requirements. Of particular interest for office applications is the relatively small but growing number of data banks that contain factual rather than bibliographic information. DIALOG users, for example, can gain rapid access to such diverse information as the output of specified manufacturing plants, the activities of philanthropic agencies and foundations, and the requirements of defense contracts.[9]

During the next decade, the number and scope of online data banks and their office application should expand significantly, especially in the important non-bibliographic sector where innovative services are enjoying an increasingly favorable reception. An online service called LegisLate, for example, provides access to a data bank of congressional legislative activity, including congressmen's voting records. Other systems offer online access to mailing lists or directories of products or services. In the United States, office use of such data banks should be stimulated by the development of so-called "videotext" information services which are now available in several European countries and Canada. Briefly, the development of these services is proceeding along two lines:

1. *Teletext* services, such as the British Broadcasting Company's Ceefax and the Didon version of the French Antiope system, employ a broadcast methodology which transmits information over the existing television network, utilizing the otherwise unused lines in the vertical blanking area of successive television frames. Thus, no new transmission frequencies must be allocated. The provider of teletext services assembles a magazine of approximately 100 pages for transmission on a continuous sequential basis at high speed. (In the British Ceefax system, the entire set of 100 pages is retransmitted every 24 seconds.) Some of the pages serve as indexes. The viewer consults the indexes, then uses a special keyboard/controller attached to the television set to display specified numbered pages.

2. *Viewdata* services, such as the Source and Compuserve in the United States, the British Broadcasting Company's Prestel, and the Canadian Telidon system, use the telephone network as the transmission medium. The resulting speed—typically 1200 bits per second—is much slower than that of the Teletext services, but the "magazine" format of preassembled pages is eliminated. The viewer first consults an index, then orders specified numbered pages transmitted. Indexes are displayed on the screen when the initial connection is established. The system first displays a broad index which is successively narrowed until a final selection of pages is reached.

Since they were developed initially for the home, videotext services have emphasized news, sports, weather, and items of general cultural interest, such as theater schedules and announcements of gallery showings. In time, however, it is expected that videotext services emphasizing data of commercial interest will develop for office applications.[10]

Computer Graphics

As computerized storage and retrieval capabilities have increased in scope and sophistication, information systems analysts have been faced with the difficult task of developing report and display formats which present

Figure 3-10. Office use of online data banks should be stimulated by the continued development of videotext information services. Such services transmit a wide range of information to specially-modified television sets. In the French Antiope system, operators utilize a hand-held controller to access specific pages of information—in this case, a national weather map. (Courtesy: John Adams Associates, Inc.)

retrieved data in a form that is easily understood and utilized by managers and others. The earliest data processing applications generated printed reports which inundated the recipient with vast quantities of data, making it extremely difficult to locate information of significance to the problem at hand. Exception reporting, in which printed reports are limited to data about those transactions or other activities which deviate from some pre-established norms, represents an improvement over unselective reports, while online systems in general, and data base management systems in particular, represent a significant advance in that they allow the user to retrieve specific items of information on demand, rather than having a lengthy report printed in anticipation of demand. In the typical online system, however, the retrieved data consists of alphabetic and/or numeric symbols which are displayed or printed in narrative or tabular form. This is the preferred display format for textual data such as listings of citations retrieved from online information banks but, if substantial quantities of statistical data are retrieved, a conventional tabular presentation can prove difficult to interpret. Following the adage that one picture is worth a thousand words, then such

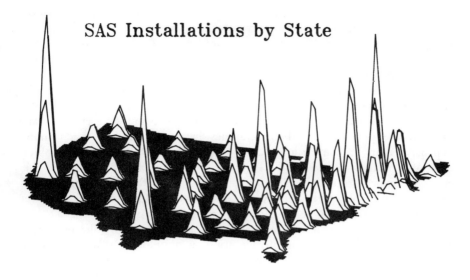

SURFACE MAP

Figure 3-11. Graphics are an increasingly important vehicle for the presentation of management information in computer-based systems. Available graphics software can generate such commonly-encountered business diagrams as pie charts, bar charts, graphs and histograms. More sophisticated graphics capabilities involve the creation of contour plots which utilize three-dimensional presentations to convey complex information. (Courtesy: SAS Institute, Inc.)

applications may be better served by a graphic presentation as an alternative or supplement to tables or narratives.

Straightforward graphic presentations involving the generation of bar charts and histograms are a commonly encountered feature in computer-based information systems. Many conventional display terminals, as previously noted, offer line-drawing capabilities that support such presentations. More sophisticated computer graphics—involving the generation of elaborate plots and cartographic displays—have been available for more than 10 years, although their use has been restricted traditionally to specialized applications in the sciences and engineering. Recent reductions in the cost of graphics-related hardware, especially graphics display terminals, and the implementation of graphics capabilities on small computers designed for home applications, have contributed to an emerging interest in graphics as an alternative to conventional means of presenting management information. Proponents contend that graphic presentations enable managers and others to communicate and assimilate information more quickly and to identify trends or deviations from pre-established norms more easily.[11] In addition, the availability of interactive graphics systems allows users to test several or many contingencies relative to a given problem and to comprehend readily the probable results of alternative courses of action. The availability of color in many graphic systems further contributes to the clarity of the presentation, especially in applications involving complex sets of variables.

At present, computer graphics are being applied to a relatively small, but growing, range of business problems. Typical applications include site location determination, market penetration analysis, and competitive analysis. While prices have fallen, graphics hardware remains much more expensive than its conventional counterpart. In 1980, prices for graphics display terminals ranged from $10,000 to upwards of $50,000, with the price of color terminals beginning at around $25,000. The price of a conventional display terminal, by way of contrast, seldom exceeds $2,000. A graphics plotter, which produces paper output, is priced in the $4,000 to $60,000 range, depending on capabilities and the need for color printing. In addition to hardware, implementation of a graphics system requires software and data. The software can be purchased in pre-written form or be developed by an organization for its own use. Several time-sharing service bureaus offer graphics software and, for many users, service bureaus represent the fastest, least expensive means of implementing a graphics system. Where service bureaus are used, the customer need only purchase or lease a graphics display terminal and provide the essential statistical data. Such data can be collected inhouse or purchased from A. C. Nielsen, R. L. Polk, or other organizations that specialize in data collection. Where cartographic displays will be utilized, additional cartographic data are required. Such data can be purchased from the U.S. Census Bureau or other sources. Taken together,

Figure 3-12. Office use of computer-generated graphics requires special terminal equipment. Left, a graphics display terminal closely resembles conventional alphanumeric CRT terminals in design. Both black-and-white and color displays are available. (Courtesy: Tektronix). Right, increased office use of graphics will generate a need for compact printing equipment capable of producing paper output. (Courtesy: Honeywell Test Instruments Division)

these implementation requirements are significantly more complex than those encountered in conventional data processing systems. It is expected, however, that interest in, and implementation of, graphic systems for management applications will be an important facet of office automation during the coming decade.

SMALL BUSINESS COMPUTERS

Background

The preceding sections of this chapter discussed the use of large-scale, time-shared computers in office applications involving the storage and retrieval of information. In such applications, the computer itself is located at some remote site, and the office gains access to it through an online terminal. This section discusses small business computers, the latest development in a continuing trend toward the placement of information-processing devices in the office.

Through the early 1970s, computer system designers emphasized the economies of scale to be derived from the use of large-capacity computing machinery serving many users in a centralized environment. The philosophical foundation for this method of managing computing resources derived from a principle called *Grosch's law* which, while never formally published, became part of the oral tradition of the computer industry. Formulated by the computer scientist Herbert R. J. Grosch in the late 1940s, it states that larger, and consequently more expensive, computers provide significantly greater information-processing power per dollar than smaller computing machines.[12] Assuming that the additional processing power is warranted, Grosch's law implies that the consolidation of an organization's computing requirements in a single centralized machine, rather than several smaller ones, results in the lowest unit cost of computing. By the mid-1960s, the development of time-sharing operating systems and telecommunication facilities for remote access made such centralization of computing resources possible. While smaller computing equipment—*minicomputers*—had been available since the late 1950s, their use through the 1960s remained limited to aerospace, industrial, and other applications requiring computing power dedicated to a specific task, such as the monitoring of a rocket flight or the control of a manufacturing process.

But the data processing industry is, if nothing else, characterized by rapid change and, by the mid-1970s, the continued viability of this historically dominant method of organizing computing resources became increasingly questioned. Two of the motives for this change are especially pertinent to office automation:

1. Many users experienced considerable difficulty in obtaining satisfactory service from centralized computing facilities. Development times for

new applications often appeared inordinately long. Job scheduling was typically beyond the user's control. Computing center personnel often seemed unresponsive to user requirements and difficulties and, when confronted by demands for improved performance, too often reverted to esoteric terminology and pseudo-technical explanations to evade accountability. As a result, many users expressed a desire for greater control over the computing resources that had become increasingly indispensable to them.

2. Coincidentally, improvements in the manufacturing of electronic circuitry—especially the replacement of discrete transistors by integrated circuits containing many electronic components on a single silicon chip—resulted in a dramatic reduction in the cost of smaller computers.[13] In 1959, for example, the DEC PDP-1, the first model in a famous family of minicomputers manufactured by Digital Equipment Corporation, cost more than $120,000. By 1968, the price of a PDP-8/I, a much more powerful successor model, had dropped to around $15,000. (Today, it sells for about $3,000.) While Grosch has on several occasions defended the continued accuracy of the law named for him, the attraction of low-cost minicomputers led many organizations to decentralize or *distribute* their computing resources, placing computer power under the users' direct control. By the mid-1970s, distributed processing had become an accepted method of computer resource management.[14]

Today, many minicomputer manufacturers and system designers direct their products to the office market. The typical office-oriented minicomputer system is configured much like a large-scale computer with disk drives, tape drives, line printers, and the ability to support online terminals for interactive applications. While prices begin at around $30,000, a fully configured system can cost upwards of a quarter of a million dollars. The most expensive minicomputer systems are difficult, if not impossible, to distinguish from large-scale computers. They offer performance characteristics which exceed those of some large-scale devices, while, at the low end of the product group, minicomputer performance is challenged by that of microcomputers.

Microcomputers

The microcomputer is the latest product of continuing improvements in the manufacturing of electronic components. *Large-scale integration* (LSI), a manufacturing technique by which many thousands of highly miniaturized circuits can be consolidated in a very small space, has resulted in a significant and continuing decrease in the cost of electronic components. The result has been another shift in the relative power of computing devices as modestly priced microcomputers can now perform work which several years ago would have required a comparatively expensive minicomputer.[15]

While the term carries the primary connotation of compact size, a microcomputer is properly defined as a computer system which incorporates a microprocessor as its central processing unit. Broadly defined, a micro-

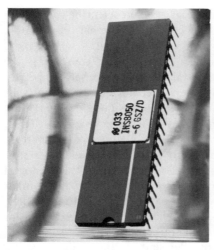 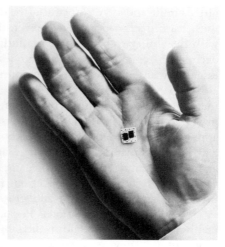

Figure 3-13. At left, the INS8050 is a single-chip microcomputer contained in a package which measures approximately 2.5 inches long. (Courtesy: National Semiconductor) At right, a memory module containing storage capacity for 8,000 characters. (Courtesy: IBM Corporation)

processor is a single-chip integrated circuit device capable of performing the arithmetic and logical operations typically associated with a central processing unit in a computer system. Microprocessors vary considerably in their capabilities and intended purposes. Microprocessors of comparatively modest power are used, for example, as controllers in digital watches, thermostats, automobile ignitions, games, and other electronic devices designed to perform functions other than those typically associated with computing or data processing. Similarly, microprocessors play an indispensable role in the operation of the interactive terminals discussed earlier in this chapter. While such microprocessors are designed for use in data processing applications and are typically more powerful than those used in the consumer products mentioned above, the terminals that contain them are not themselves microcomputers. Lacking resident computing power, they cannot be programmed by the user, but are, instead, dependent on interaction with an external computing device. The absence of user-programmability is likewise a limitation of the text-editing systems described in Chapter Two. For purposes of this discussion, user-programmability and a resulting generality of application distinguish microcomputers from such pre-programmed, specialized information-processing equipment.

Microcomputer manufacturers typically distinguish three broad markets for their equipment: scientific computing, personal computing, and business data processing. Scientific computing is characterized by complex, often repetitive calculations performed on relatively modest amounts of data. While the input, output, and storage requirements of such applications are often simple and straightforward, the device selected for scientific computing must feature a broad range of powerful and efficient mathematical operators with the ability to perform computations accurate to as many as 10 or more decimal places. Microcomputers designed for scientific computing are essentially an outgrowth of programmable calculators and other computing aids which have long proven indispensable in technical applications. An increasing number of microcomputers, however, are designed for the rapidly growing personal/household computing market in which automation is applied to such varied tasks as checkbook balancing, income tax preparation, and portfolio planning; the storage and retrieval of recipes and addresses; and the activation and control of lamps, burglar alarms, and other electrical appliances. Recreational applications involving games, quizzes, and similar diversions are an important and growing facet of this computing market, as are educational applications. Personal/household computers vary considerably in sophistication and in the range of peripheral equipment supported. The simplest devices consist of a controller and keyboard designed for attachment to a customer's own television set and tape recorder. In many cases, however, personal computers can be expanded, through the addition of peripherals and software, into systems suitable for office applications.

Figure 3-14. Microcomputer systems represent the next logical step in the continuing evolution of computing equipment. Designed for office operation, the typical microcomputer system features floppy disk storage and will support one or more online terminals and related peripherals. In keeping with the idea of enhancing white collar productivity through the provision of tools, there has been considerable office interest in personal computers designed for use by managers, scientists, engineers and other so-called "knowledge" workers. (Courtesy: Three Rivers Computer Corporation)

While scientific and personal computing applications are important, most microcomputers are designed for office data processing and are described collectively as small business computers. The most commonly encountered elements in a microcomputer system designed for office operations are described in the remaining sections of this chapter.

Equipment Configurations

Regardless of size, a computer system consists of a combination of a central processor, memory circuits, and peripheral equipment. The central processor in a microcomputer system is, as previously noted, a microprocessor—a computing device composed of a large number of digital circuits integrated on a single, small silicon chip. Most microcomputer system developers do not manufacture their own microprocessors but, instead, purchase them from such companies as Intel, Zilog, AMI, TI, Motorola, Mostek, and Synertek. These microprocessors differ somewhat in design and capabilities. While many of the differences are technical and do not affect their performance in the most commonly encountered office applications, certain features do have an impact on equipment selection and consequently require further discussion. These features include word-length, memory-type and size, and bus structure.

As applied to computer architecture, the term *word* denotes the number of bits of information which a central processing unit can access or manipulate in a single operation. *Bits,* as described in the preceding section on computer terminals, are the binary elements used to encode information for computer processing—the information, in this case, being either instructions that the computer will execute or the data which are the objects of those instructions. Typically, the longer the word, the more flexible and powerful the computing device. The central processing units of large-scale computers, for example, are generally capable of accessing 32 bits at one time, while most minicomputers access 16-bit words. By way of contrast, the earliest microprocessors were four-bit devices. Such microprocessors are still used in calculators and similar numerically oriented, special-purpose machines, but they are too limited for use in microcomputers intended for business applications. Unlike numbers, the alphabetic characters encountered in much office data and computer programs are each encoded using a group of eight bits, which are referred to collectively as a *byte.* To access a given character, a four-bit microprocessor requires two operations, thus significantly reducing processing speed. Most small business computers overcome this limitation by employing an 8-bit microprocessor, and a small but increasing number employ 16-bit microprocessors for more powerful processing capabilities. In addition to their speed advantage, 16-bit microprocessors generally respond to a more extensive instruction set than 8-bit devices, thus permitting programs to be shorter and more efficient.

Like the microprocessor, the small business computer's internal memory consists of integrated circuits. Programs and data under immediate execution usually are stored in *random-access memory* (RAM), sometimes called user memory or working memory. RAM circuits often are supplied in increments of 2K or more bytes, where 1K bytes equals 1,024 characters. Most microcomputers are equipped initially with 8K to 16K bytes of random-access memory which is expandable to 64K, 128K, or more bytes. Many small business computer applications require a minimum of 32K to 64K bytes of random access memory to accommodate programs and data in their entirety. Smaller memory capacities degrade processing speed by forcing increased reliance on the slower secondary storage devices discussed later in this chapter.

Random-access memory circuits are constructed from semiconductor materials that are volatile—that is, their content is erased whenever the machine is turned off or the power supply is otherwise interrupted. Thus, programs and data are stored typically on secondary magnetic media, such as tapes or disks, and are read into random-access memory at the start of the work day or when otherwise required. For system monitors and other frequently used programs, however, *read-only memory* (ROM) circuits provide a convenient form of non-volatile storage. Read-only memory consists of integrated circuits which contain programs or data pre-written by the circuits' manufacturer to specifications provided by the microcomputer system developer or customer. As noted in the previous chapter's discussion of text-editing systems, programs stored in ROM circuits are described as *firmware* to distinguish them from the *software* stored in random-access memory or on magnetic media. As the name implies, the content of read-only memory can be used but not overwritten or erased. However, several variant forms—such as programmable read-only memory (PROM) and erasable programmable read-only memory (EPROM)—do permit modifications under specified conditions.

In terms of the interrelationship of components, small business computers can be divided into two broad groups: 1) those that consist of a microprocessor configured with pre-selected memory circuits and peripheral devices offered for sale as an integrated system; and 2) those that allow the customer to select from a wide range of available hardware, some of which may be developed by companies other than the original microcomputer vendor. The latter approach is analogous to the purchase of a stereo system by separately selecting individual components. As is true of stereo component systems, successful implementation depends on compatibility, which in turn, is influenced by bus design. Used in this context, the term *bus* denotes the network of connecting wires along which data and instructions, in the form of electrical signals, travel between the microprocessor, memory, and peripheral devices. The most widely used microcomputer bus design is called the *S-100 bus*. It is designed for use with Intel 8080 and equivalent microprocessors, such as the popular Zilog Z80, and has become a *de facto* microcomputer standard. A wide range of memory circuits and other components are available for attachment to the S-100 bus, giving those systems that utilize it an advantage in applications where customized expansion is required. It should be noted, however, that several of the most popular microcomputers—notably the Radio Shack TRS-80 series and the Apple II and III—do not use the S-100 bus. However, with regard to those computers, utilization of a proprietary bus design is not a significant limitation since the large number of Radio Shack and Apple installations has attracted companies who manufacture circuitry and components specifically for them. This is not invariably the case with other non-S-100 systems.

Peripheral Devices

The peripherals in a computer system consist of three broad groups of devices—input equipment, output equipment, and secondary storage. As is true of most small business computer systems and their large-scale counterparts, the most commonly encountered input device is a keyboard which is attached physically to an output device such as a display terminal. When configured in this way, the resulting peripheral device is identical with the online display terminals described earlier in this chapter. For the production of hardcopy output, most small business computer systems will support one or more printers. The most commonly encountered are either impact printers or a non-impact (thermal) device which generates individual characters from a dot matrix pattern. They are analogous to the Receive-Only (RO) teleprinters described in a preceding section. For applications in which the small business computer will be used for word processing, typewriter quality output can be obtained with either a daisywheel-type Qume or Diablo printer or with an NEC Spinwriter, a printer that uses a thimble-shaped printing element to generate fully formed, high-quality characters. The requirements of most small business computers can be satisfied with some combination of the above equipment. For special applications, some microcomputers will support additional input/output devices such as image digitizers, pen plotters, and voice synthesizers.

As previously noted, the capacity of random-access memory within a given microcomputer is necessarily limited, and RAM circuits are themselves volatile. The user's requirement for prolonged storage of data and programs consequently must be met by secondary storage media and devices that retain data in machine-readable form pending later use.

While magnetic tape cassettes are used in some small business computer systems, the dominant form of secondary storage consists of floppy disks. As discussed in Chapter Two, a floppy disk is a circular-shaped piece of polyester coated with a magnetic recording material. Two sizes are available: the standard floppy disk, which measures eight inches in diameter, and the mini-floppy, which measures 5.25 inches. Floppy disk size has an obvious impact on storage capacity, but capacity is influenced additionally by the

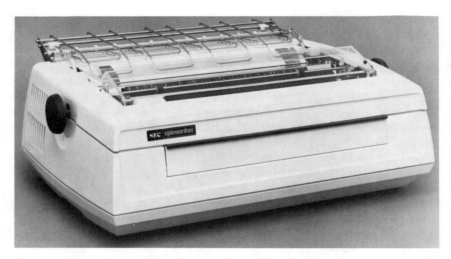

Figure 3-15. An increasing number of small business computers combine data processing with word processing. Where typewriter-quality output is required, the NEC Spinwriter uses interchangeable thimble-shaped printing elements to generate paper documents that are equivalent in quality to those created by the daisywheel printers provided with text-editing systems. (Courtesy: NEC Information Systems)

recording method employed. Depending on the system, data may be recorded on one or both sides of each disk. Most newer systems employ dual-sided recording. As is true of hard disks, the recording surface itself is divided into a series of concentric tracks, each of which is further subdivided into sectors. Within each sector, storage capacity is determined by recording density. The single-density format, originally developed by IBM and subsequently adopted by other manufacturers, divides the standard-size floppy disk into 77 tracks, each of which is itself sub-divided into 26 sectors. Each sector will store a maximum of 128 characters for a nominal total capacity of 256K characters per recording surface. However, an increasing number of microcomputer systems will support a double-density disk drive which records 256 characters per sector for a total capacity of 512K characters per recording surface. Dual-sided disks effectively double these capacities to 512K characters per single-density disk and 1,024K characters per double-density disk. Thus, a microcomputer system equipped with two standard-size floppy disk drives could have online access to over two million characters of data and/or instructions. Mini-floppy disks, being smaller, can contain between 71K and 179K characters per recording surface. A small business computer system equipped with two mini-floppy disk drives can provide online access to as many as 358K characters of data and/or instructions.

Where standard or mini-floppy disks provide insufficient online storage capacity, some small business computer systems will support a conventional hard-surface disk drive designed to accept removable disk packs. More often, however, a Winchester-style disk drive is used. The Winchester drive features a fixed hard-surface disk capable of storing as many as 100 million characters of data and/or instructions. Some small business computers will support multiple Winchester drives for total online storage capacities in excess of several hundred million characters. In addition to greater storage capacity when compared to floppy disks, Winchester disks offer more operating speeds. When compared to conventional, removable disk drives, the Winchester-style drive is less complex and consequently more reliable. As is true of conventional hard surface disks, small business systems employing Winchester disks must be provided with back-up protection in the event of disk failure. This back-up capability typically relies on the periodic recording of data onto floppy disks or magnetic tape cartridges.

Microcomputer Software

The preceding sections discussed small business computers in terms of equipment or hardware. The term *software* is used to denote the programs or sets of instructions which enable computers to accomplish specific tasks. These programs may be prepared by the computer system vendor, by the user, or by some third party. With regard to the data on which it operates, small business computer software typically is stored on magnetic media—tapes or disks—until required, at which time all or portions of a given program are read into random access memory. Certain frequently executed programs may be maintained permanently, as previously noted, in read-only memory, where they are continuously available. These programs—which, as previously discussed, are called firmware to distinguish them from conventional software—frequently are written by the microcomputer vendor and delivered with the equipment.[16]

Firmware is an especially useful technique for implementing that group of programs called *system software*. Historically, the computer industry has distinguished system software—those programs which enable the computer to function and control its own operations—from *application software* (those programs which perform some user-specified task). The most important category of microcomputer system software is a group of supervisory programs called the *operating system*. These programs are designed to accept and respond to user-oriented commands to load and execute other programs. For continuous availability, the operating system may be implemented in firmware and maintained in read-only memory. Whether implemented in firmware or software, the operating system programs are written typically by the microcomputer equipment vendor. Alternatively, some third parties have developed operating systems for specific types of microcomputers. In some cases, these independently developed operating systems are preferable to those provided by the equipment vendors. An example of such a software product is *CP/M,* a widely used operating system developed by Digital Research Incorporated and subsequently implemented on many different microcomputers.

In addition to an operating system, most small business computers are provided with utility programs designed to facilitate the user's own software development efforts. One of the most important of these utility programs is an *editor.* Not to be confused with the text-editing programs provided for word processing applications, the editor allows programs to be conveniently entered into online storage and modified as required. Other useful utility programs assist in the debugging of programs, control input and output devices, and perform commonly encountered information-processing tasks such as sorting.

User-programmability is the feature that distinguishes microcomputers from the other microprocessor-controlled information-processing machines discussed elsewhere in this book. All of the available small business computers will support user programming in one or more of the higher-level languages—BASIC being the most widely available for this purpose. It, however, rarely is implemented in exactly the same way on any two systems. Some lower-priced microcomputers offer a comparatively primitive version of BASIC, which limits the programmer to a few straightforward calculations involving integers. Others provide a more powerful BASIC interpreter which permits floating point calculations, the manipulation of character strings, and similar features which extend the utility of the system for office applications. Unfortunately, the various available versions of extended BASIC are largely incompatible with one another.

Besides BASIC which, being easy to learn, is well suited to inexperienced users, some small business computer systems can be programmed in versions of FORTRAN, COBOL, Pascal, or APL. In some cases, special user-oriented programming languages have been developed for implementation on specific microcomputer systems. These languages usually are patterned after BASIC and are designed for ease of learning and use by persons without extensive computing experience.

While small business computer systems are user-programmable, many users prefer to purchase pre-written application software rather than write programs themselves. Such software, the nature and availability of which varies significantly from system to system, is commonly obtainable from one or more of the following sources:

1. Many microcomputer system developers offer one or more prewritten program packages designed to enable users to implement such commonly encountered business applications as general ledger maintenance, accounts payable, accounts receivable, and inventory management immediately. In most cases, the cost of the packages compares very favorably with the cost of programming labor, and speed of implementation will result in earlier

realization of cost savings and productivity improvements in those applications in which the microcomputer system replaces a labor-intensive manual procedure.

2. Some small business computer systems consist of a pre-selected combination of microcomputer equipment and pre-written application software designed for the automation of activities within specific industries such as manufacturing or pharmaceuticals. When marketed in this manner, the resulting configuration is called a *turnkey* system. The name suggests that an office need only add its own data, turn on the key, and begin immediately realizing the benefits of automation. In reality, successful implementation may require significant customer involvement and accommodation, but properly designed turnkey systems can free users of many of the problems associated with separate equipment procurement and software development.

3. In some cases, third parties—persons or companies other than the microcomputer system vendor—have developed application programs for implementation on particular systems. In most cases, these programs are developed by service bureaus or software companies, but some microcomputer vendors have large, active user groups whose members contribute software to a library maintained and publicized by the equipment vendor. In some cases, software companies develop programs for a particular type of microprocessor. Such programs generally can be executed by any small business computer equipped with that microprocessor. Many programs are developed, for example, for systems that utilize the Intel 8080 or equivalent microprocessors and the S-100 bus structure discussed earlier in this chapter.

As indicated above, much pre-written software is designed for the automation of the most commonly encountered financial applications. In addition, a growing number of microcomputers are equipped with word processing software. In such cases, the microcomputer system physically resembles the display-oriented text-editing systems described in Chapter Two. Such systems differ, however, from conventional text-editing systems in their greater reliance on operator-memorized codes rather than the more convenient dedicated keys provided with equipment designed specifically for word processing. In addition, microcomputer-based word processing software packages may not provide the advanced editing capabilities found on the most sophisticated text-editing systems. On the other hand, where word processing requirements are straightforward and occasional, the microcomputer-based approach offers the advantage of a single, multi-function piece of equipment that gives the user both word and data processing capabilities for the price of a stand-alone display-oriented text-editing system.

Intelligent Terminals

As more and more office applications employ computers of various types and sizes, there will be an increased need to communicate data electronically between such devices. Some small business computers can communicate with a compatible small business system or with a larger computer. Such communication capabilities are important, for example, in those applications where the results of local information processing must be made available to one or more remote computers, or where local computing or online storage capabilities must be supplemented by those of a larger computer. In some cases, large data files are maintained on remote computers and, using telecommunications, are copied selectively or "down-loaded" for local processing by the small business computer.

With most small business computer systems, telecommunications capability is implemented via an optional software package and an RS-232C interface that allows the microcomputer to be connected to an external modem or acoustic coupler. When configured in this manner, the small business computer functions as an intelligent terminal—a device which, as previously described, is capable of local information processing and interaction with a remote computer. The particular telecommunications protocol observed varies from system to system. In most cases, the communicating microcomputer emulates the asynchronous ASCII terminals described earlier in this chapter. Some small business computer systems, however, are capable of higher speed bisynchronous communication.

Procurement Considerations

Small business computers differ from other types of information-processing equipment in that they increasingly can be purchased in stores where the customer has the opportunity to observe and try several different models prior to making a procurement decision. Such stores, which are located in most medium- to large-size cities, may be either independently owned or franchised. They often sell microcomputers and related support components developed by smaller companies which lack direct marketing capabilities. Such companies often have the most interesting and sophisticated products. Recognizing the potential of the small business market, however, microcomputer system developers increasingly are opening their own retail outlets for direct sale to end users. Radio Shack was the first company to take this approach, selling its TRS Series microcomputers through its large network of retail outlets and, in large cities, through Radio Shack Computer Centers. Recently, other vendors, such as IBM, Xerox, and Digital Equipment Corporation, have begun opening computer stores for direct sale to the general public.

Prices for small business computers, as with other types of automated office systems described in this book, are necessarily subject to change. At the time this chapter was written, the nominal price range for microcomputer-based systems was $2,000 to $20,000, with most workable configurations designed for office applications falling in the $7,000 to $15,000 range, including the cost of software and essential peripherals. Assuming both a

4-year useful life and the purchase of an annual maintenance contract averaging 10 to 15 percent of total hardware value, the approximate annual equivalent cost of a typical system ranges from $2,500 to $5,000. This is approximately 20 to 40 percent of the annual total of wages and fringe benefits paid to a clerical employee performing tasks that such a computer could perform. Viewed in this light, small business computers offer considerable potential for cost-effective operation in a wide range of office applications.

4
Micrographics

Source document microfilming • Computer-output microfilm • Computer-assisted retrieval (CAR)

Micrographics denotes that field of information processing that is concerned with techniques associated with the production, handling, and use of microforms—photographic information storage media which contain images recorded in greatly reduced size.[1] Examples of specific types of microforms include roll microfilm, microfilm cartridges, microfilm jackets, microfiche (film cards), aperture cards, fiche-size card jackets, and tabulating-size jacket cards. The images they contain are properly called *microimages,* and they may be created by filming paper source documents or from machine-readable, computer-processable data via computer-output microfilm (COM) technology. Their content may consist of either textual or graphic information.

As briefly noted in Chapter One, micrographics provides an example of an established technology that has assumed increased significance as a component in the "paperless information systems" which characterize much office automation.[2] The preceding chapter discussed the importance of machine-readable, computer-processable data as an alternative to paper documents in office applications; computer-based information systems, however, are not necessarily paperless. In batch-processing applications, computer programs often are designed to produce voluminous, multiple copy paper reports, while many online systems utilize printing terminals to generate hardcopy output. Even when display terminals are used, online systems rarely replicate the entire information content of source documents. In most applications, the complete conversion of source document content to machine-readable form cannot be justified from the standpoint of data entry labor and/or online storage costs. In addition, the content of certain types of documents—those containing graphic rather than textual information are an obvious example—can prove technically and economically difficult to convert to, and maintain in, machine-readable form. Furthermore, some

Figure 4-1. Microforms—photographic information carriers containing images recorded in greatly reduced size—are available in a variety of roll and flat formats.

applications require the preservation of both document content and appearance. Whether textual or graphic in nature, source documents, or photographic images of them, sometimes must be retained for legal reasons, even if their information content is entirely replicated in machine-readable data bases.

Yet, the continued maintenance of large quantities of information in paper form poses significant problems for the office. The storage of paper records requires considerable amounts of space—an office resource that is expensive now and is likely to prove even more costly in the years ahead. While a systematic record destruction program can minimize storage space requirements for those documents that can be discarded shortly after creation, records which require long-term retention demand a continuing commitment of valuable floor space. Similarly, paper documents can be time consuming and expensive to disseminate. With the exception of single documents transmitted via facsimile technology in a manner to be discussed in Chapter Six, paper files cannot be accessed easily from remote locations, and their bulk and associated maintenance problems typically prohibit their wholesale replication and distribution to multiple offices. Whether in one office location or several, the development of effective paper filing systems is a very difficult task, and there are few successful examples of automated paper filing systems. Most of the available devices are merely motorized files which offer a performance advantage over conventional filing equipment in only a narrow range of office applications.

Many of the problems described above can be addressed successfully through micrographics. Recognized since the 1920s as a cost-effective vehicle for the long-term storage of inactive business records and still widely associated with that activity, micrographics technology has played an important role in the management of active office records for over 20 years. The development of self-threading cartridges, microfilm jackets, and aperture cards in the late 1950s and early 1960s demonstrated the superior performance advantages of microforms, when compared to paper documents, in a wide range of office applications. Since the 1960s, micrographics systems analysts and equipment manufacturers have increasingly emphasized the information retrieval and dissemination advantages of microforms, and much recent attention has been given to the interfacing of micrographics with computers and other office technologies.

Whether used alone or in conjunction with other technologies, micrographics offers considerable potential for productivity improvements in office information processing. Most obviously, microforms conserve office space, an important capital resource in the calculation of total productivity, as discussed in Chapter One. Less obvious is the role of computer-output microfilm and computer-assisted microfilm retrieval in conserving valuable computer resources—notably, processing time and online storage. In terms of labor savings, micrographics contributes to improved management and clerical productivity through the more rapid retrieval of information essential to work performance. In certain applications, the use of microforms can further reduce the amount of clerical labor required for the establishment and maintenance of paper files. The remaining sections of this chapter discuss recent developments in micrographics technology in terms of their significance for office automation. The subsections that immediately follow focus on three important facets of source document microfilming: (1) the development of a new generation of microprocessor-controlled microfilm equipment which offers significant performance advantages when compared to older devices; (2) the emergence of new recording technologies designed specifically for office applications; and (3) the concept of small office microfilm (SOM), which extends the range of micrographics to new groups of users. Later subsections emphasize the importance of the computer/ micrographics interface in office applications.

SOURCE DOCUMENT MICROFILMING

"Intelligent" Microfilm Equipment

In office applications, operability by nontechnical personnel is critical to the acceptance and successful implementation of information-processing systems. Historically, micrographics equipment manufacturers have followed the general trend in information processing by emphasizing increasingly sophisticated capabilities while retaining simplicity of operation. Through the mid-1970s, microfilm cameras and other equipment designed for source document applications made extensive use of prevailing mechanical technology. Such equipment—much of which remains in production and widespread use—performs effectively, but lacks the flexibility offered by some of the newer equipment utilizing microprocessors. Given the rapid change that has characterized office work in recent years, the ability to respond to new application requirements is of the greatest importance. Recognizing this, micrographics equipment manufacturers have begun to take advantage of the developments in electronics technology discussed in the preceding chapter. The result has been a new generation of "intelligent" microfilm products which derive their operating characteristics from integrated circuitry—specifically, microprocessors, read-only memories, and random-access memories. While mechanical components remain important, the increased reliance on electronics has contributed to improved reliability, simplified maintenance, more compact equipment design, and greater energy efficiency. The most important advantages, however, are enhanced performance characteristics and an increased flexibility which provides the user with a greater measure of protection against equipment obsolescence. These advantages are seen clearly in two micrographics product groups: source document cameras and microfilm retrieval units.

Source document microfilm cameras are divided customarily into three groups by method of operation and application. Rotary cameras, which take their name from the high-speed transport mechanism that moves documents past a lens and a light source where they are recorded on 16-mm microfilm, have been used widely in applications where work throughput is a paramount consideration and the source documents to be filmed are of uniformly good quality. Planetary cameras, by way of contrast, microfilm source documents that have been positioned individually on a flat surface. As is true of rotary cameras, roll microfilm is produced, but the documents remain stationary during exposure. The resulting image quality is generally markedly superior to that obtainable with rotary cameras. Planetary cameras are thus preferred in roll film applications involving source documents of varied quality, but the necessity of positioning documents individually results in significantly degraded work throughput. While the throughput of rotary camera operators may average one thousand or more documents per hour, the typical planetary camera operator will only attain half that speed. A third type of source document microfilmer, the step-and-repeat camera, is planetary in design, but produces microfiche rather than roll microfilm.

While manually operated planetary and step-and-repeat cameras offer high image quality and, in the case of step-and-repeat cameras, microfiche output, the degraded throughput associated with them is incompatible with automated office operations which emphasize improvements in labor productivity. Over the last few years, however, several micrographics equipment manufacturers have introduced devices which combine the operating speed of rotary cameras with the ability to produce microimages of a quality normally associated with planetary cameras. Operating under microprocessor control, these automatic-feed planetary and step-and-repeat cameras transport documents at high speed to an exposure surface where they are stopped momentarily and microfilmed. Using special pre-formatted cards inserted

Figure 4-2. Automatic-feed planetary cameras offer both high work throughput capability and excellent image quality. The microprocessor-controlled Bell and Howell ABR-100 responds to specially designed program documents. (Courtesy: Bell and Howell)

into the source document stream, the operator can activate one or several programs stored in read-only memories in order, for example, to initiate two-sided microfilming operations, to interrupt the microfilming process, or to activate or alter the machine's film-indexing capabilities.

Further performance enhancement and increased flexibility result from the application of electronics to the design of step-and-repeat cameras, traditionally the most expensive yet inflexible type of microfilmer. The conventional step-and-repeat device features a fixed film movement mechanism which creates fiche with images arranged in a predetermined pattern of rows and columns. The customer typically is offered a choice of several or many standard and/or nonstandard fiche formats and reductions, but the choice must be made at the time the camera is purchased. Conversion from one format to another, when possible at all, often involves a complete replacement of the camera's film transport mechanism and lens system. An office with a requirement for two fiche formats—24X, 98-frame and 42X, 208-frame, for example—must purchase two cameras. Several newer step-and-repeat cameras, however, employ integrated circuits to overcome this limitation. With regard to these cameras, film movement and image positioning are controlled by microprocessors which respond to any of several programs stored in read-only memories. To change recording parameters, the operator uses a control panel to select the desired fiche format and, if required, changes the lens. Since additional or replacement programs can be incorporated in the camera at any time, the customer is afforded considerable obsolescence protection. Given the attractiveness of microfiche in office applications, this improvement in step-and-repeat camera design is of great significance.

The preceding discussion briefly mentioned the role of microprocessors in the initiation and control of microfilm indexing. Conventional rotary and planetary cameras can be modified to place small opaque marks called *blips* below each image on 16-mm microfilm. The resulting blip-encoded images can be retrieved using specially designed readers and reader-printers equipped with electronic circuitry capable of counting the blips and stopping at a frame number entered at an operator keyboard. As discussed later in this chapter, blip encoding is often used in conjunction with computer-assisted retrieval. While effective in the straightforward retrieval procedure described above, the recent development of microprocessor-controlled microfilm cameras and retrieval units has enhanced both the variety and applicability of blip encoding. In addition to the standard blip adjacent to each film image, several newer microfilm cameras will mark images with blips of several sizes, or position blips above as well as below all or selected images on film. These multi-level blips are especially useful in applications where documents are processed in batches. In such applications, which often are encountered in office information processing, the large or top blips can be used to mark the first document in each batch, while smaller or bottom blips can be recorded beneath each subsequent document. As previously noted, some newer microfilm cameras will respond to pre-formatted instruction cards that activate the appropriate blip pattern based on programs stored in read-only memories. To retrieve a document image numbered within a given batch, several microprocessor-controlled readers and reader-printers are capable of counting blips in several sizes or positions on film. These

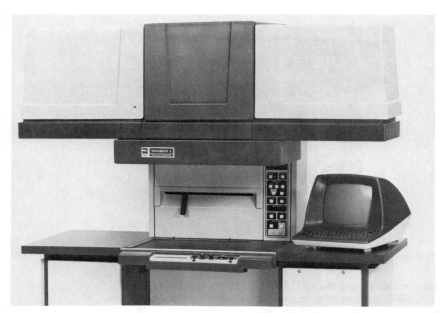

Figure 4-3. The TDC microprocessor-controlled DocuMate I is one of a new generation of step-and-repeat cameras. (Courtesy: Terminal Data Corporation)

"intelligent" microfilm retrieval units also may be pre-programmed to display and/or print specified sequences of frames automatically. Some models provide a user-accessible random-access memory for the storage of frame numbers entered manually or, in computer-assisted retrieval applications, electronically.

New Recording Technologies

Whether of conventional design or microprocessor-controlled, most source document cameras record latent microimages on silver halide film stock. Following exposure, the film customarily is developed, using a combination of chemical solutions and water, in an external processor. The advantages of the silver halide process account for its historical and continued dominance of photography in general and of microphotography in particular. Its performance characteristics are well understood; it can produce useable microimages from source documents of varied quality, and it is the only recording medium compatible with existing standards for the production of archival quality microforms. But even its strongest proponents acknowledge

Figure 4-4. Designed for CAR or other retrieval applications, the VISCO controller is a microprocessor-based device which converts a roll-film reader or reader-printer into an automated retrieval unit. (Courtesy: Visual Systems Corporation)

that the silver halide process has two limitations that are significant in office applications, especially those involving active, growing files:

1. Silver halide images must be exposed and developed in separate steps. Wet chemical processing is, at best, inconvenient and, at worst, requires more attention to technical details than can reasonably be expected of clerically trained office personnel.

2. Even though unexposed areas may remain on a given roll of film or fiche, a processed silver halide microform cannot be re-exposed to record additional document images.

Neither of these obstacles is insurmountable. The perceived inconvenience of external wet chemical-processing office applications can often be minimized or eliminated through the use of a camera-processor, a device which exposes and develops microimages in one continuous operation. Some, however, may not sufficiently wash the film to meet archival-processing standards. Available silver halide camera-processors can produce fully developed roll microfilm, microfiche, and aperture cards. They are operated in much the same manner as office copiers and feature pre-measured containers of required processing chemistry which minimize any potential inconvenience associated with supply replenishment. Similarly, the inability to add new images to an existing film or fiche can be overcome by using microfilm jackets, jacket cards, and card jackets. Alternatively, however, the office user can now select one of the several available microfilm cameras that employ nonsilver recording technologies. These devices, which are intended specifically for office applications, are designed to address one or both of the limitations of silver halide technology noted above.

Introduced in 1977, the 3M 1050 step-and-repeat camera is the first commercially available device capable of recording source document images on *dry silver* microfilm. Dry silver microfilms, which have been used successfully in computer-output microfilmers since the late 1960s, are exposed to light but are developed by heat alone without wet chemicals. The dry silver process also is used in microfilm reader-printers to make paper enlargements of microimages. Its attractiveness in office applications is obvious. The 3M 1050 can convert office documents to microimages on fiche in one continuous operation. The film processor is contained within the microfilmer and accepts exposed dry silver film directly from the camera section without operator intervention. Developed fiche are delivered precut

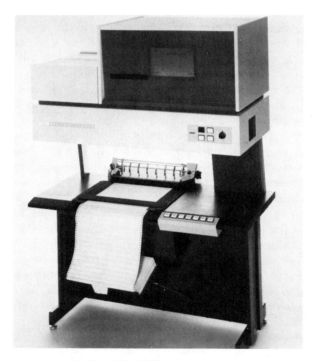

Figure 4-5. The 3M 1050 step-and-repeat camera is the first commercially available source-document microfilmer to utilize dry-silver microfilm. Being thermally processed without the aid of chemical solutions, dry-silver microfilms are well suited to office applications. (Courtesy: 3M Company)

and ready for viewing or duplication. Special site preparation, plumbing, and provisions for the disposal of chemical wastes are not required. The dry process, however, does not meet current archival standards criteria.[3]

The dry silver process overcomes potential difficulties associated with wet chemical development in office applications but, like silver halide microfilm, dry silver film loses its sensitivity following exposure and processing. While microforms offer significant advantages in space savings, retrieval, and file maintenance, the growth of office-based micrographics applications has been impeded historically by difficulties in accommodating growing files—that is, files to which source documents must be added regularly. The need to control the space required for the maintenance of such files and to facilitate their retrieval is a commonly encountered office requirement in such diverse fields as insurance, medicine, criminal justice, engineering, and banking. As previously noted, the requirements of such applications often can be accommodated by using microfilm jackets or related microfilm carriers, but the creation of such microforms is necessarily a labor-intensive activity involving a series of separate work steps and equipment. Updatable microfiche systems achieve the same result—the ability to accommodate file updates in an active office environment—with a single piece of equipment and simplified work steps. At the time this book was written, two updatable microfiche systems—the System 200 Record Processor, marketed by A. B. Dick Systems, and the Microx System II, marketed by the Microfilm Products Division of Bell and Howell—were in office use.[4]

The two systems consist of a step-and-repeat camera-processor which exposes and develops microimages in one continuous operation. They differ, however, in recording technology and capabilities. The System 200 Record Processor employs a variant form of electrostatic recording called transparent electrophotography. Latent images, created by a pattern of electrostatic charges on a film surface, are developed through the application of toner, in the manner of an office copier. New document images can be added to empty spaces on previously exposed fiche by merely reinserting the fiche into the camera and indicating the frame to be exposed. Images cannot be erased once recorded, but they can be obliterated, double-exposed, or overprinted with an additional message. The Microx System II uses photoplastic film on which latent microimages are formed from a pattern of electrostatic charges. The images are developed by deforming the film through the application of heat, and they subsequently are fixed by cooling. As is true of the System 200, new images can be added to empty spaces on previously exposed fiche. Unlike the System 200, however, the Microx System II can erase previously recorded images and, if desired, replace them with new images.

In mid-1980, A. B. Dick Systems acquired the marketing rights to a third updatable microfiche system called the MicrOvonics File. Originally

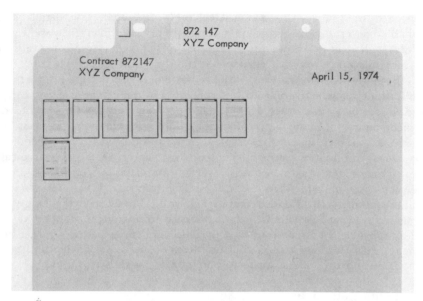

Figure 4-6. Using transparent electrophotography, the A.B. Dick System 200 records source document images on updatable microfiche. The fiche itself is shaped like a miniature office file. (Courtesy: A.B. Dick Systems)

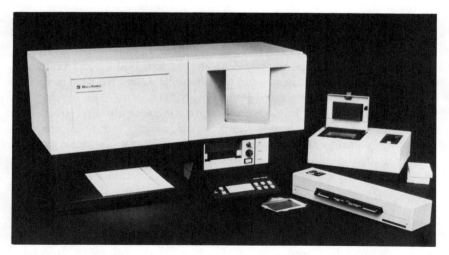

Figure 4-7. The Microx System II is an updatable microfiche system which records source document images on photoplastic film. A planetary-style, desk-top unit, the Microx camera includes a monitor for the immediate verification of recorded images. (Courtesy: Bell and Howell)

developed by Energy Conversion Devices Incorporated, few details are available concerning its recording technology or performance capabilities. While it has been shown in prototype versions on several occasions, production models were not available at the time this book was written.

Both the System 200 and Microx System II are designed for office applications where active files will be converted to microfiche at a very early point in the files' life cycle and where additional documents will be recorded on fiche on a continuous basis. In terms of office automation, they can be viewed as another manifestation of a trend toward the development of "paperless information systems" noted in Chapter One. In a typical application, source documents are converted to updatable fiche shortly after their creation or receipt in the office. The resulting fiche files can be maintained manually or under computer control. When a request for a given file is received, the updatable fiche is removed from storage, duplicated on a desk-top device designed for that purpose, and immediately returned to storage, thus eliminating potential problems of file control. The duplicate fiche is given to the requestor for use in a reader or reader-printer.

Small Office Microfilm

The micrographics products described in the preceding sections, as well as those discussed later in this chapter, are comparatively high-priced devices designed for large-scale office applications. At the time this chapter was written, microprocessor-controlled step-and-repeat cameras, for example, usually were priced in the $20,000 to $35,000 range, with some devices costing as much as $100,000. Planetary cameras are less expensive, but when the cost of processors, duplicators, and related support equipment is considered, the price of a source document microfilm production facility routinely exceeds $30,000 and often is considerably higher. Microprocessor-controlled retrieval units are priced at around $10,000 to $15,000 each, and an updatable microfiche system, with duplicators and required peripheral equipment, costs between $20,000 and $40,000.

The micrographics industry is, however, increasingly interested in the development of products which effectively address the special information-processing requirements and constraints of small offices. The acronym *SOM* often is used to denote this Small Office Microfilm as a concept, a market, and a product group of significance to office automation.[5] SOM products extend the range of micrographics technology to those offices that cannot afford or effectively utilize complex systems. The SOM market includes both small autonomous organizations and individual departments or other small operating units within large organizations. SOM thus can be viewed as the micrographics equivalent of the distributed information-processing configuration previously discussed in Chapter Three. As is true of micro-computers, SOM products are designed for office users who lack micrographics expertise and experience and who must work within budgetary constraints of varying severity. The SOM product group includes cameras, processors, duplicators, production support equipment, readers, and reader–printers—all designed with economy and convenience as foremost considerations. In the early 1970s, it was thought that an entire SOM system could be made available at a cost less than one thousand dollars. Today, $10,000 is a more realistic estimate of the required initial capital investment.

Simple, economical, but effective cameras are critical to full microform utilization in the small office. The most attractive and commonly encountered SOM cameras are of the planetary type. Typically priced at $1,700 to $4,500 and designed for desk-top operation by clerical personnel, the SOM planetary produces 16-mm roll microfilm which, following processing, usually is inserted into cartridges or microfilm jackets. SOM applications often rely on service bureaus or other external laboratories for the processing of exposed microfilm. While several low-volume, table-top processors are available, most SOM users prefer to avoid the semi-technical worksteps involved in the installation and operation of a microfilm processor and associated quality-control equipment. For applications where the use of external processing services is prohibited or impractical, several manufacturers have developed desk-top camera-processors which produce fully developed 16-mm microfilm. One such device—the EOM 6100, developed by Electro-Optical Mechanisms—delivers the processed film in pre-cut strips ready for insertion in microfilm jackets.

Several table-top rotary cameras were available in the $3,000 to $6,000 price range at the time this chapter was written. These devices are necessarily less versatile than the more expensive high-volume rotary microfilmers utilized in many large-scale office applications, but they do offer recording capabilities and operating characteristics that are well suited to the comparatively straightforward requirements of most SOM applications. There are, unfortunately, no comparably priced step-and-repeat cameras, although, as discussed below, there are a number of available examples of excellent, low-cost microfiche readers and duplicators. In the absence of low-cost fiche production capabilities, SOM applications requiring flat microforms typically rely on microfilm jackets and card jackets, both of which can utilize support equipment designed for microfiche.

Desk-top diazo and vesicular microfiche duplicators constitute one of the most attractive and well-developed groups of SOM products. Unlike highly automated, production-type duplicators which are designed for high-volume centralized microfilming operations, desk-top SOM duplicators are the microfiche equivalent of convenience copiers. Priced at $1,000 to $5,000, they are intended for use in decentralized installations where single copies of selected fiche will be made on demand. Supply costs average 5 to 10 cents per fiche. Since each fiche may contain as many as 98 pages, the

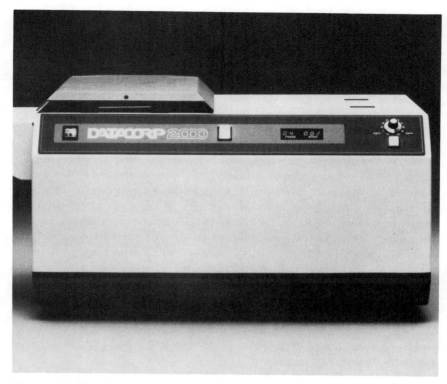

Figure 4-8. Simple, economical cameras are critical to full microform utilization in the small office. The Datacorp 2000, developed by Electro-Optical Mechanisms, is a camera-processor specifically designed for operation by non-technical personnel in office applications. Copier-like in design and size, it produces fully processed microimages on pre-cut strips ready for insertion in microfilm jackets or other microform carriers. (Courtesy: Datacorp)

cost of duplication is significantly lower than that of an equivalent quantity of paper copies. In addition to microfiche, most desk-top duplicators will accept microfilm jackets, card jackets, and other fiche-size flat microforms.

In a typical application, a desk-top duplicator is used to produce dissemination copies of master microfiche maintained in a noncirculating file, thus minimizing the potential for loss or other problems of file integrity. As noted in the previous discussion of updateable microfiche systems, this use of desk-top duplicators is not limited to SOM applications. They can be employed effectively in large-scale applications where demand duplication is required.

It is a rare office, large or small, that cannot afford a microfiche reader. With prices in the $200 to $300 range, fiche readers are unquestionably the

Figure 4-9. Readers and reader-printers are the means whereby users interface with micrographics systems. A microform reader (left) projects microimages on a screen for viewing at or near full size. A reader-printer (right) provides selective hardcopy enlargements of displayed microimages. (Courtesy: Micron Corporation)

least expensive type of micrographics equipment, and in terms of cost/performance characteristics, they are among the most attractive of all information-processing products. The attractiveness of newly introduced models is continually increasing, in terms of both general engineering and attention to human factors. Commonly encountered features include interchangeable lenses, modular design, a choice of screen sizes, and high-quality optics designed to generate bright, clear images suitable for display in high ambient light. Several of the readers designed for COM applications closely resemble the display-type computer terminals described in Chapter Three.

For applications requiring selective hardcopy enlargements of displayed microimages, several microfiche reader-printers are available in the $1,500 price range. Designed for office applications, they are operated in much the same manner as convenience copiers. They include many of the most desirable features of microfiche readers combined with the ability to make paper prints.

COMPUTER-OUTPUT MICROFILM

COM Technology

The phrase "computer/micrographics interface," introduced earlier in this chapter, denotes the combined application of computers and micrographics to information-processing problems that cannot be addressed effectively by using either technology alone. The increasing popularity of COM is the most obvious and pervasive example of the computer/micrographics interface.[6] As noted in the introduction to this chapter, the execution of many computer programs results in the production of voluminous, multiple-copy paper reports. The acronym COM denotes the product (computer-output microfilm), the process (computer-output microfilming), and the device (computer-output microfilmer) which converts machine-readable, computer-processible digital data to human-readable textual or graphic information on microfilm without first creating paper documents. A computer-output microfilmer, or COM recorder, is a computer peripheral device and a high-speed microfilmer capable of printing data at speeds of two to five pages per second. As a computer peripheral, it is capable of either online or offline operation. Online COM recorders are connected to the computer on which the data to be recorded are processed. They typically are located in data processing centers, and their microform output is manually delivered to the office in the manner of paper reports. Offline recorders are stand-alone devices that accept digital data on an appropriately formatted magnetic tape. The tape, resulting from the execution of a computer program, must be carried physically to the COM recorder where it is mounted on an integral or attached tape drive. The tape may be generated by any one of a wide range of computers, and the recorder can be located anywhere. Offline recorders are used widely by service bureaus and are one of the most important sources of COM-generated microforms in office applications. Alternatively, an inhouse, offline recorder may be operated by an organization's micrographics department, data processing facility, or records management group. Many offline COM recorders incorporate a small minicomputer which is used to reformat tapes containing data originally prepared for line printers. The availability of such minicomputer-controlled COM recorders has greatly simplified and facilitated the transition from paper output to microforms by virtually eliminating the need to modify existing application programs. As a result, most office applications can be converted from hardcopy printouts to microfilm or microfiche without significant data processing effort.

With offline recorders priced in the $60,000 to $150,000 range exclusive of support equipment, at the time this chapter was written, the typical COM installation was centralized and intended to serve many offices as a shared resource. Several of the less expensive devices are actually COM recorder/processors. They are designed to produce fully developed microfiche

Figure 4-10. A COM recorder is both a computer peripheral device and a high-speed microfilmer designed to record computer-processed data on microfilm or fiche without an intervening paper copy. The DatagraphiX AutoCOM II is a COM recorder-processor that delivers fully developed, cut fiche in a single operation. (Courtesy: DatagraphiX, Inc.)

ready for immediate viewing and/or duplication. If archival permanence is a requirement, these devices may not meet archival-processing standards. They can be installed successfully in offices where recording volumes are high enough to permit full utilization.

Regardless of the mode of operation or machine location, the COM recorder is designed to convert machine-readable digital input into human-readable information on microforms. The input itself consists of information to be recorded, as well as coded instructions which direct the recorder. With the exception of equipment designed for special applications, available COM recorders use one of two microfilming methodologies: cathode ray tube (CRT) photography or laser beam recording. CRT photography is the oldest and most prevalent form of COM recording. Data from magnetic tape, or, in the case of online COM recorders, from the computer itself, is displayed as a page of information on the screen of a cathode ray tube located inside the COM recorder. The CRT display then is photographed by a high-speed microfilm camera, the film is advanced, the display is erased, a new page is displayed, and the process is repeated. The recording medium is silver-halide microfilm which, in most cases, is developed in an external processor.

In laser-beam COM recording, a low-power helium-neon laser is used to activate and expose points in a dot matrix pattern on film. Lasers can be directed with great precision at very high speeds. When the laser scans a matrix on film, the beam exposes only those points necessary to shape individual characters. Unlike the light generated by a conventional CRT display, laser light is concentrated and sufficiently powerful to expose unconventional photographic media. The available laser beam COM recorders all utilize thermally processed silver microfilm, a light-sensitive recording medium that is developed by the application of heat. The use of thermally processed silver microfilm in source document applications was discussed briefly in a preceding section of this chapter. In COM applications, the thermal silver processor can be attached directly to the COM recorder, thus permitting the exposure and development of microfilm or microfiche in one continuous operation. Because thermally processed silver microfilm is developed without wet chemicals, laser beam recorders have proven especially popular in those applications where the COM recorder will be located in the data processing facility. As discussed later in this book, laser beams also are used in certain xerographic printers which are becoming increasingly popular in data processing and word processing applications.

To emulate report production on pre-printed business forms, all COM recorders will merge digital data with a reproduction of a static format. This normally is accomplished with a form slide, a transparent piece of glass bearing the image of a business form with blank spaces to be filled in with data by the COM recorder. The form slide is superimposed on the data during recording. Some COM recorders can store several form slides and interchange them automatically, as required, within the same microform. As an alternative to form slides, several COM recorders store form definitions in internal memory and generate them with the appropriate data, under program control.

Regardless of the recording method utilized, the COM-generated microform must be processed and duplicated. The possibility of internal processing was mentioned in the foregoing discussion of thermally processed silver COM recorders and silver halide COM recorder-processors. As of this writing, most COM recorders, however, utilize silver halide microfilm and external processing. Following processing, vesicular or diazo duplication typically is used to produce the required number of copies.

In terms of output capabilities, most available COM recorders will produce both 16-mm and 35-mm roll microfilm and standard 105- by 148-

mm microfiche. Several newer COM recorders, reflecting changes in the popularity of the various microforms, will produce only microfiche. Given the wide selection of reasonably priced, high-quality microfiche readers, it is understandable that fiche is the dominant microform in COM applications. The use of 35-mm COM-generated microfilm typically is limited to applications involving computer-produced engineering drawings, charts, maps, circuit diagrams, or other graphic presentations. These images are created by sophisticated high-level graphic COM recorders. Most COM recorders are, however, alphanumeric devices. Designed for the most commonly encountered business applications, they are unable to handle graphic output. Some essentially alphanumeric recorders, however, are equipped with limited "business graphics" capability. For example, they can generate line charts, bar charts, pie charts, labelled axes, histograms, and other presentations encountered in management reports and other office documents. The previous chapter briefly discussed the projected growth of such graphics in management-oriented computer applications.

COM as Line Printer Alternative

Since the late 1960s, COM recorders have been used as high-speed, paperless replacements for line printers in applications requiring the timely production of voluminous printed reports from machine-readable data. While there are several types of available line printers, the most widely used models are impact-type output devices featuring a printing chain consisting of characters represented on embossed metal slugs linked in an endless loop. The chain rotates horizontally to bring characters into their appropriate print positions. A bank of hammers behind each print position is selectively activated to drive a sheet of paper into an inked ribbon and the chain, thus printing the characters. Since several hammers may be activated simultaneously, the device appears to print entire lines at one time (hence the name). Other line printers use drums rather than chains, but their operating characteristics are similar. Despite recent advances in printing technology—notably, the development of non-impact page printers discussed later in this chapter—line printers remain the most widely used paper output device in batch-processed computer applications.

In terms of report production, COM recorders offer significant throughput advantages when compared to line printers. The rated or maximum speeds of line printers have increased over the years to the current high of approximately 3,000 lines per minute. Actual speeds depend on several factors, including the size of the printable character set, and typically prove significantly lower than rated speeds. The most widely used line printers operate at speeds ranging from 900 to 1,500 lines per minute. Multiple copies generally are produced on four- or five-ply interleaved carbon form sets. For copies in excess of the form set size, the print program must be re-executed. This combination of low actual printing speed and multiple copy production via carbon paper can result in excessively long printing times for reports intended for distribution to many users. The production of 20 copies of a 200-page report easily, for example, could require one hour of printing plus additional time for the decollating, bursting, binding, and distribution of output.

By way of contrast, COM recorder-rated speeds range from 7,500 to 50,000 lines per minute. Although actual speeds vary with page lengths and the method of input preparation, throughput rates of 15,000 to 20,000 lines per minute are realistic for the faster recorders. Thus, the production of the COM master of a 200-page report on a single 42X microfiche would require less than 5 minutes. Allowing additional time for microfilm processing and diazo or vesicular duplication, the required 20 copies can be ready for distribution within half an hour. With its greater speed, a single COM recorder offers output capacity equivalent to perhaps 10 line printers—without a ten-fold increase in cost. In replacing line printers with COM recorders, computing centers and computing service bureaus translate this increased recording speed into cost savings which, in turn, can be passed on to their office-based customers in the form of lower output charges. It is now widely recognized that, when compared to line printers, COM offers potential for substantial cost reduction in the production of long, multiple-copy reports that are updated on a frequent schedule. In such applications, cost reductions of 40 to 50 percent are attainable. In addition to these economies of production, COM offers certain economies of use that can prove significant in office applications. The storage space required for paper reports, for example, will be reduced. Conversion to COM-generated microfilm or microfiche can speed the retrieval of information when compared to the awkward handling often associated with bulky paper reports. In applications requiring long-distance distribution of voluminous reports, conversion to COM can reduce the cost of mailing and associated handling significantly.

In addition to the potential for cost reduction in certain applications, COM offers the additional advantages of improved quality and versatility when compared to line printer output. In terms of output quality, the fourth and fifth copies in a line printer-generated carbon interleaved form set are often of marginal utility. All COM duplicates are made directly from the master microform and are, consequently, of uniform quality. With respect to output versatility, most line printers are equipped with a standard print chain of about 64 characters, including the upper-case alphabet, numeric digits, and selected punctuation symbols. While extended print chains are available—an upper/lowercase print chain is a commonly encountered example—any expansion of the printable character set necessarily results in a reduction in printing speed because it will take a longer time for any given character to reach its intended print position on a line. Thus, a printer rated at 1,100 lines per minute with a 64-character print chain

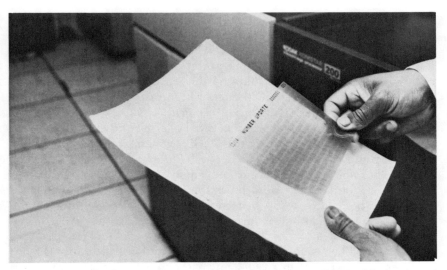

Figure 4-11. COM technology successfully addresses the problems posed by the computer's ability to produce voluminous quantities of printed information. Above, a single microfiche containing 208 pages occupies less than one-sixth the area of a single sheet of computer printout paper. (Courtesy: Eastman Kodak Company)

may print only about 550 lines per minute when lower case capability is added. This speed reduction, combined with the limitations of output quality noted above, have restricted the use of line printers in computer-based word processing and similar applications.

The output capabilities and characteristics of most COM recorders are not, however, so restrictive. While the basic COM character set totals 60 to 70 characters, most alphanumeric recorders offer an optional extended set of 90 to 128 characters, including the lower-case alphabet and additional special punctuation, foreign, and mathematic symbols. In addition, most COM recorder manufacturers will create special characters to customer order. The expansion of the COM character set is, furthermore, not accompanied by a degradation of output speed, as is true of line printers.

While the simplest alphanumeric COM recorders, like their line printer counterparts, can print characters in only one size and intensity, several recorders offer the user a choice of type styles. In some cases, characters of varying sizes and intensities can be intermixed within a given frame or line on film or fiche. Thus, for example, the body of a report can be printed in a regular type font; section headings can be printed in bold type and selected words or lines can be recorded in italics. Where exceptional typographic quality and versatility are required, some recorders such as the Information

International COMp80/2 offer output capabilities equivalent to the most sophisticated photocomposition devices.

In terms of page formatting, COM recorders generally have emulated line printers in preserving the appearance of the familiar 11- by 14-inch computer printout page containing 64 lines of up to 132 characters each. Thus, the most widely used COM microfiche formats provide the equivalent of 208 computer printout size pages per fiche at $42\times$ reduction and the equivalent of 270 computer printout size pages per fiche at the standard $48\times$ reduction. While the 11- by 14-inch page format may be suitable for data processing applications, it is not well suited to word processing and related office applications. Fortunately, most COM recorders can produce other output formats. For word processing applications, for example, line lengths can be reduced to 80 characters, thus producing the equivalent of an 8.5- by 11-inch document image and increasing the number of images per fiche to 325 at $42\times$, or 420 at $48\times$. At $24\times$, a reduction that is an NMA standard but is used rarely in COM applications, the 8.5 by 11-inch document image results in a COM microfiche format that is identical with the prevailing NMA standard for microfiche of source documents. As an alternative to conventional formatting in applications where a premium is placed on increased recording capacity, the spaces between frames can be eliminated entirely, producing a "scrolling" effect in which the data appear to be recorded as a series of continuous lines on fiche or film.

COM and Page Printers

As briefly noted in the preceding section, advances in printing technology have resulted in the development of a new generation of high-speed, paper-oriented computer output devices which are designed to replace line printers in centralized, high-volume applications.[7] Referred to as page printers to reflect the fact that their operating speeds are measured in pages rather than lines, these new devices rely on nonimpact copier technology rather than the typewriter-derivative technology characteristic of impact printers. The two most popular models—the Xerox 9700 and the IBM 3800—utilize xerographic recording, which is discussed more fully in the next chapter. Because they are based on existing copier/duplicators—the Xerox 9700, for example, is based on the Xerox 9400 duplicator—these page printers are sometimes described as "intelligent copiers," although that term is better applied to a somewhat different type of equipment described in Chapter Five. Unlike conventional copiers, the page printers accept machine-readable, computer-processible input rather than paper documents. A laser beam is used to create latent images of individual characters out of a matrix-type pattern of electrostatic charges on a selenium drum or other intermediate surface. A toner is applied, and the resulting developed images transferred to an ordinary sheet of paper. Operating at speeds in the 18,000 line per minute range, these non-impact printers can

111

Figure 4-12. A combination of lasers and xerographic technology has resulted in the development of a new generation of high-speed, paper-oriented computer-output devices, designed to replace line printers in centralized, high-volume applications. The Xerox 9700 produces highly legible, letter-size copies at two pages per second. (Courtesy: Xerox Corporation)

produce 2 pages per second. They thus overcome the speed limitations of available line printers. In addition, they produce highly legible output on plain letter-size paper that is more conveniently handled than its prevalent 11- by 14-inch line printer counterpart. Because individual characters are generated by a laser from pre-stored definitions, these page printers are much more versatile than line printers. The Xerox 9700, for example, can produce characters in a wide range of type fonts and sizes. As is true of COM recorders, form slides can be used to eliminate the need for specially designed paper stock.

There are other significant similarities between these page printers and COM recorders. Both, for example, are non-impact recording devices. Both operate at speeds in excess of 15,000 lines per minute. Laser recording is common to both technologies. The interrelationship of the two technologies is reflected in the fact that the Xerox 9700 can be ordered with a microfiche production subsystem that consists of an integral COM recorder. Page printers and COM recorders address the essential limitations of line printers, described in the preceding section. In certain applications, both enjoy a significant economic advantage as an alternative to line printer production of paper reports. But page printers and COM recorders differ in one important respect—page printers produce paper output, while COM recorders produce microforms. Even though the available page printers produce letter-size output that is more compact and somewhat easier to handle than 11- by 14-inch printouts, they do not address the fundamental problems of storage and dissemination associated with paper reports. In addition, as non-impact devices, page printers are incapable of making carbon copies. Print programs must be executed repeatedly to produce the required number of multiple copies. While the repeated printings occur at very high speed, COM recorders retain a production advantage in applications involving long reports that are distributed to many users on a frequent basis, although COM's advantage may not be as marked as it is against line printers.

COM and Online Systems

The preceding sections discussed the role of COM as an alternative to paper printers in report production applications. Chapter Three indicated, however, the growing importance of online systems in office applications. Given the anticipated reductions in the cost of online storage discussed in that chapter, information systems analysts have given considerable thought and attention to the role of COM in an online environment. It is possible to view COM in any of three such roles: (1) as an alternative to online systems; (2) as a complement to online systems; or (3) as a supplement to online systems.

The view that COM represents an alternative to online systems in certain applications is based on a distinction between online systems in general and that subset of online systems which are designed to satisfy a requirement for real-time information. The term *online* properly denotes a system configuration characterized by interconnected equipment, while the term *real-time* refers to a method of processing information, although they are often used interchangeably.[8] An online information-retrieval system, for example, is one in which information stored on disk drives or other direct access devices is accessed via terminals. The terminals and the storage devices physically are connected to the computer on which the information is processed. Thus, the opposite of online is offline.

A real-time system, by way of contrast, is one in which information about a transaction or other event is processed by a computer at the time the event occurs. This is the case, for example, with regard to airline reservation systems. The opposite of real-time processing is *batch* processing in which information reflecting the occurrence of a given transaction or other event is collected and processed, with information about similar transactions, at regularly scheduled intervals some time after the transaction takes place. Real-time information systems necessarily operate online, but some online systems provide access to information that is processed in batches. Alternatively, many batch-oriented systems rely on printed reports to convey information about previously processed transactions.

In office applications, there is an obvious and direct relationship between the accuracy of information and its value. In transaction-oriented applications, the accuracy of information depends on how current it is. Real-time systems, as defined above, provide access to information about the latest transactions. Where such information is required, there is no alternative to a real-time system. Many office applications, however, can be satisfied with access to information that is current within the last business day. Such applications typically involve batch processing, but, given the relative slowness of line printers, computer centers and service bureaus are not always able to guarantee report production on a daily basis, especially when the reports are long and many copies are required. The user may consequently be forced into an online system. While the cost of such systems, as noted in Chapter Three, is decreasing, it is hardly negligible. In many applications, the production of printed reports is the more economical alternative. COM recorders, with their greater work throughput capabilities, can enable computing centers and service bureaus to produce long, multiple-copy reports on a frequent schedule, thus eliminating the need for an online system in applications where access to real-time information is not required. Where greater currency of information is required within a batch-processing application, COM recorders can be used to produce reports at more frequent intervals than is possible with line printers. Batch-processed reports can be produced, for example, several times daily, thus providing the office user with information that is no more than a few hours old.

While this combination of batch processing with COM-generated reports is well suited to a wide range of office applications, real-time information systems sometimes are required. Such systems, as previously discussed, require online components, but COM can play an important, complementary role in the implementation of a real-time system. Consider, for example, a computer-based text-editing system in which reports which are subject to revision are maintained conveniently on disks or other direct access storage media. Users responsible for the revisions can make them at online terminals. When all revisions have been completed, or after some predetermined period of time, the reports can be converted to document images, via COM, on microform. If desired, an abbreviated version of the report—an abstract or other summary, for example—can remain online with an indication of the film or fiche location of the full text. Such a hybrid online/COM system can prove more economical than a total online system in that the disk storage requirement is reduced significantly. While disk storage costs, as reported in Chapter Three, have decreased steadily over the past 10 years, they are still nowhere near as low as microfilm storage costs for an equivalent quantity of data.

Finally, assessing COM as a supplement to online systems, it is important to note that online systems are vulnerable to hardware or software malfunctions which can result in the partial or complete unavailability of data. COM-generated printouts can serve as a relatively inexpensive form of back-up information in the event of online system failure. Similarly, COM-generated printouts can play a useful role in applications where limitations on available direct access storage or other factors do not permit the continuous availability of data online. Some data bases are, for example, online for only specified hours during a business day. A COM-generated report on film can provide access to information during the remaining periods.

Computer-Input Microfilm (CIM)

Computer-input microfilm is a variant form of optical character recognition. A CIM device scans, interprets, and translates human-readable information on microfilm into machine-readable digital data on magnetic tape or other media. Thus, CIM is conceptually the opposite of COM, and the input to a CIM device can be COM-generated film or fiche. In anticipation of CIM applications, most COM recorders generate individual characters in the OCR-B-type font, a sans serif-type face that is designed specifically for applications involving optical character recognition.

The attractiveness of CIM is obvious. Tapes, disks, and other forms of magnetic storage are more expensive than microfilm, are much bulkier, and are vulnerable to deterioration over time. Microfilm is economical, compact, and, when properly processed and stored, archivally stable. Information on microfilm offers the additional advantage of being human-readable, thus permitting the simultaneous creation of reports for reading and computer re-entry. While the advantages of microfilm over paper in optical character recognition systems have long been recognized, CIM technology still is developing and its likely future impact on office operations is unknown.

At the time this book was written, available CIM systems consisted of two expensive, special purpose devices: (1) FOSDIC, a computer-input microfilm that has been in use at the United States Census Bureau, in various versions, since the early 1950s; and (2) Grafix I, a complex and powerful CIM system developed by Information International Incorporated. Grafix I uses a unique combination of hardware and software to read microfilmed source documents or COM-generated microimages created in a variety of fonts and formats. Although any 35-mm microfilm can be accepted, best results reportedly are obtained from high-quality COM-generated images prepared by graphic recorders such as the Information International FR-80.[9]

COMPUTER-ASSISTED RETRIEVAL (CAR)

The CAR Concept

Computer-assisted retrieval is an information-processing methodology that combines the space saving and other advantages of microform storage with the ability of computers to rapidly manipulate index data. CAR systems use computers and online storage to establish, maintain, and search an index keyed to document images which reside in a separate microform file. From the computer standpoint, the CAR approach simplifies data entry and online

Figure 4-13. In the CAR configuration shown here, the microprocessor-controlled retrieval unit—a Kodak IMT-150—is linked to the online terminal through a specially designed CAR interface. (Courtesy: Eastman Kodak Company)

storage by limiting those activities to index data rather than entire documents. Microfilming documents is a much less expensive procedure than keystroking or otherwise converting their information content to machine-readable form. Even with recent increases in the cost of silver, microfilm storage—as previously noted—remains cheaper than magnetic media. From the standpoint of micrographics systems design, CAR extends the range of micrographics to office applications which cannot be accommodated successfully by conventional retrieval approaches.[10]

With the possible exception of those applications that emphasize the long-term storage of infrequently referenced documents, a well-designed micrographics system must provide the office user with a convenient means of identifying and locating desired microimages. In applications involving well-organized collections of office documents, retrieval requirements often can be accommodated successfully by straightforward techniques which involve little or no special equipment. But where individual documents are not, or cannot be, microfilmed in a readily useable sequence, or where individual images must be located and displayed very rapidly, the conventional, predominantly manual retrieval approaches typically prove ineffective. While the CAR concept has attracted much recent attention within the context of office automation, the idea of automated support for microimage retrieval actually dates from the mid-1940s. During the 1950s and early 1960s, a number of experimental systems were developed for use in applications requiring the rapid retrieval of scientific and technical documentation. By the mid-1960s, several working systems were available for sale, and **ADSTAR** (Automated Document Storage and Retrieval) was developed to denote this emerging technology. In the United States, for example, Eastman Kodak developed a group of retrieval products—the most famous being marketed under the brand name Miracode®—which featured index data encoded on microfilm in binary form, adjacent to the document images to which they pertain. Despite limitations, these systems played an important role in information retrieval at a time when computers were much less powerful and much more expensive than they are today.

The first true CAR applications date from the early 1960s, perhaps the most famous examples being the Walnut document storage and retrieval system developed by the Central Intelligence Agency, and the microfiche retrieval subsystem developed for Project Intrex at the Massachusetts Institute of Technology. These and similar "one of a kind" systems relied on customized components, but by the early 1970s, "off the shelf" products suitable for use in CAR applications became available. These products included microfiche or aperture card retrieval devices and 16-mm microfilm reader-printers capable of accepting film coded with image count marks or "blips." At nearly the same time, several complete CAR systems, consisting of an integrated configuration of hardware and software, were introduced for sale on a "turnkey" basis.

Despite these promising beginnings, widespread interest in, and acceptance of, CAR concepts failed to develop until the late 1970s. Today CAR systems are regarded as an automated alternative to manual filing in office applications requiring the management of paper documents. The application of CAR systems, however, is not limited to source documents. They can facilitate the retrieval of COM-generated images as well. The remainder of this chapter discusses the technology of computer-assisted retrieval in terms of the characteristics of hardware and software that are suitable for office applications.

The Micrographics Subsystem

A computer-assisted retrieval system consists of interrelated computer and micrographics subsystems. The main feature of each subsystem is a data base consisting either of machine-readable index data recorded on computer-processible magnetic media or of document images in microform. Within these subsystems, hardware and software components are used to establish, maintain, and search the data bases. In terms of hardware, each subsystem includes input, output, and data base storage devices. In some cases, as discussed later in this chapter, selected components from each subsystem may be linked together electronically.

In the micrographics subsystem, the input device may be either a source document camera or a COM recorder. The output device is typically a reader or a reader-printer. The storage device is a manual or automated file. The storage medium on which the microform data base is maintained is usually

16-mm microfilm or 105- by 148-mm microfiche, although other microforms can be utilized. Some CAR systems include additional components designed to transmit digitized microimages to remote video monitors or facsimile printers. Such systems are discussed briefly in Chapter Six.

Historically, CAR applications have emphasized the roll microforms, especially 16-mm microfilm cartridges and cassettes. In such applications, documents can be filmed using virtually any camera capable of blip encoding. As used in micrographics retrieval, the term *blip* denotes a small rectangular mark that is recorded with all or selected document images. The blip typically appears in the lower portion of the film frame, beneath the document. The blips, which are identical to one another, have no intrinsic information content. At retrieval time, specially equipped readers and reader-printers will count them to advance the film to an operator-specified frame, in a manner described later in this chapter. Consequently, blips are sometimes called "image count marks." Blips can be placed on 16-mm film by any one of a number of available rotary and planetary microfilm cameras. In COM applications, blips can be generated via a form slide. Blip-counting readers and reader-printers suitable for 16-mm cartridge and cassette applications are available from several vendors. The typical device features a calculator-style keyboard into which the operator enters the number of the desired frame. The reader-printer's logical circuitry counts the blips, stopping the film at the indicated frame. Prices for such equipment ranged from $9,000 to $13,000 at the time this chapter was written. Simpler units, priced at around $5,000, substitute a light-emitting-diode (LED) display/counter for the keyboard. The operator advances the film until the desired frame number is displayed on the counter. Several vendors recently have introduced microprocessor-controlled retrieval units which are pre-programmed to: (1) count several types of blips; (2) store multiple frame locations for more convenient searching; and (3) perform other microfilm search operations that are not possible with most conventional equipment. The performance characteristics of these devices were described in an earlier section of this chapter.

In many CAR applications, the blip-counting reader or reader-printer operates as an offline, stand-alone device independent of other system components. Following a search of the computerized index, the operator must select the appropriate cartridge or cassette manually and mount it on the retrieval unit. When not in use, cartridges and cassettes typically are stored in drawers, file cabinets, or similar containers. One system, the Ragen MRS-95, includes a special retrieval terminal designed to store 300 microfilm cartridges in an interior carousel. The retrieval terminal operates online to a mini-computer which transmits cartridge and frame numbers directly to it. The terminal automatically selects the indicated cartridges, queues them up as required, and then displays successive frames. Other blip-counting retrieval units, including the models mentioned earlier in this section, can

Figure 4-14. Random-access files for microfiche and aperture cards, such as the Access System M, provide both automated retrieval and misfile protection. (Courtesy: Access Corporation)

be equipped optionally with an electronic interface which allows the device to operate online to a computer. This interface converts the reader or reader-printer into a special-purpose, receive-only terminal. The numbers of pertinent frames are delivered to the reader or reader-printer by the computer, thus eliminating manual key entry. The operator, however, still must manually select and mount the appropriate cartridges or cassettes.

A similar approach is used in CAR systems employing microfiche recording of document images. The computer-maintained index is first searched to determine the fiche and frame locations of relevant microimages. In some applications, the fiche themselves are stored in conventional trays, drawers, or panel files. Alternatively, there are several automated approaches to microfiche storage and retrieval. Fiche can be stored, for example, in power files. A more complex group of products provides true random access to stored fiche. These systems house each fiche in an envelope-like carrier that is notched with an identifying number. As many as several hundred thousand carriers can be accommodated by some systems. Within the file itself, individual carriers are stored in random order. To retrieve a given fiche, the operator enters its number at a keyboard attached to the file. The file's search mechanism scans the notched carriers, locating the indicated fiche and ejecting it from the file. The operator then must remove the retrieved fiche from its carrier and take it to a reader or reader-printer. When display or printing activity is completed, the fiche can be reinserted in the file at any point. As random-access files, these systems provide misfile protection in addition to automated retrieval. In most cases, they can operate with microfilm jackets, aperture cards, and other flat microforms as well as microfiche.

Figure 4-15. The Ragen 95, a turnkey CAR system, utilizes a specially designed retrieval unit to eliminate operator handling of cartridges. The cartridges, each containing approximately 4,500 images, are automatically selected on instructions received from the system's minicomputer. (Courtesy: Ragen Information Systems)

Another group of automated microfiche display systems provide storage, retrieval, display, and printing capabilities in a single unit. These devices utilize special carousels or cartridges to house individual fiche and retrieve them on instructions entered at an operator keyboard. The retrieved fiche and designated frames are displayed automatically on a reader screen. Printing capability is provided as well. In most cases, the fiche are attached to special notched clips or housed in specially marked carriers to facilitate identification and retrieval. As is true of the microfilm reader-printers described above, these automated microfiche retrieval units are capable of online operation through an optional CAR interface which converts the unit to a receive-only terminal. When operating online, these retrieval units receive fiche and/or frame numbers directly from the computer on which the index data are maintained, thus eliminating operator entry of those numbers.

The Computer Subsystem

The computer subsystem within a CAR system consists of a combination of hardware and software designed to establish, maintain, and process an index which provides access to microimages relevant to particular retrieval requests. The required computer hardware includes four groups of components: (1) a central processing unit; (2) one or more input devices; (3) one or more output devices; and (4) devices and media for the storage of the index data itself. The exact hardware configuration will necessarily vary from application to application, depending on retrieval requirements and the method of system implementation.

CAR systems can be implemented on either a customized or a turnkey basis. The latter approach provides a pre-configured system of computer and micrographics components purchased as a single product. The computer

Figure 4-16. Turnkey CAR systems consist of a pre-selected configuration of hardware and software designed for specific micrographics retrieval applications. The 3M Micrapoint System features a microcomputer with pre-written CAR software linked to a microfilm reader-printer. The system is designed for office applications. (Courtesy: 3M Company)

subsystem includes preselected hardware and pre-written software. In terms of their central processing units, turnkey CAR systems typically rely on minicomputers or, in the case of smaller systems, microcomputers. Such devices, as discussed in Chapter Three, offer extremely attractive cost/performance characteristics and are well suited to the requirements of turnkey CAR applications. Whether a minicomputer or a microcomputer is used, the central processor in a turnkey system is typically dedicated to micrographics retrieval and cannot be used for other data processing tasks. While the central processor may be programmable in its native state, the user does not have access to those capabilities.

Customized—that is, user-developed—CAR systems may utilize either a time-shared or a dedicated central processor. In time-sharing applications, the central processor may be operated by an inhouse computing center or by a service bureau. It may be either a large-scale computer or a minicomputer. It can be used, of course, for other data processing and information-retrieval tasks. Customized CAR systems also can utilize a dedicated minicomputer or microcomputer. As described in Chapter Three, a number of small business computers have cost/performance characteristics that are well suited to CAR applications.

As is true of the central processing unit, the computer subsystem's input/output and index storage configuration will vary with the method of implementation. The typical CAR application operates in an online mode. Index data are key entered via a terminal and transmitted to a computer which organizes and maintains them on disks or other direct access storage media. Some turnkey systems provide a special index entry workstation for this purpose. These workstations include a microfilm camera which permits simultaneous index entry and filming. In user-developed CAR systems, however, index data usually are entered via an online display or printing terminal, although an offline key-to-tape unit can be used. The same online terminal that is used for index entry, or one identical to it, may be used for retrieval operations. Such operations involve the operator entry of one or more commands followed by a computer response which is either displayed on a screen or printed onto paper. Many users prefer a display terminal for the input of index data and a printing terminal for the recording of computer output during interactive index searching.

The particular terminal selected will necessarily vary with the requirements of the computer used as the central processor in a given CAR system. Most applications utilize a conventional ASCII terminal of the type described in Chapter Three. Most turnkey CAR systems provide such a terminal in a configuration that is hardwired directly to a dedicated central processor. Where a remote, time-shared computer is used, the terminal is connected to the telephone network via a modem or an acoustic coupler. With regard to terminals to be used for the entry of index data, many systems utilize a "smart" CRT device capable of displaying a formatted screen—that is, a screen with labelled fields bearing the names of each of the index parameters relevant to a given document image. The data entry clerk then enters the appropriate index values in the blank areas adjacent to each of these labelled fields.

Once converted to machine-readable form through terminal entry, the index data are stored on a direct-access storage medium. Disks are used most widely for this purpose, although the type of disk utilized will vary with the computer configuration selected and the quantity of index data to be stored. Where a large-scale computer or minicomputer is utilized as the central processor in a CAR system, the direct access storage medium will be some type of hard disk. Even the smallest minicomputer systems provide hard disk capacity in excess of 10 million characters, and disk capacities in excess of 100 million characters are not unusual. Large-scale computers typically have sufficient disk storage capacity to accommodate even the largest CAR applications.

Inadequate disk storage capacity is, however, a frequent impediment to the use of microcomputers in CAR systems. As discussed in the preceding chapter, microcomputers typically are equipped with floppy disk storage in capacities ranging from 250,000 to slightly more than one million characters. Capacities approaching 2.5 million characters are provided with a few systems, but storage requirements in excess of those amounts must be accommodated by removing one or more recorded disks to offline storage and replacing them with empty disks. An increasing number of microcomputer systems, however, will support a Winchester-type hard disk drive which extends the system's online storage capacity to 10 million or more characters. Alternatively, some microcomputer systems can be configured as intelligent terminals with communication capabilities which allow them to draw on the storage facilities of remote large-scale computers to supplement local floppy disk capacity.

Indexing Software

It is relatively easy to select appropriate hardware components for the implementation of a CAR system. CAR software is, however, quite different. For purposes of this discussion, CAR software denotes the programs which a computer executes to establish, maintain, search, or otherwise process the machine-readable index to the CAR system's microform data base. The micrographics equipment in a CAR system does not require software in the conventional sense. While, as noted at the beginning of this chapter, an increasing number of micrographics products are microprocessor-controlled with programs stored in read-only memory (ROM) circuits, such programs are pre-written by the equipment manufacturers and are not the customer's responsibility.

Computer software suitable for CAR applications can be developed or otherwise obtained by: (1) writing programs for a specific CAR application and equipment configuration; (2) purchasing pre-written programs; or (3) purchasing a turnkey system which includes pre-written software as a component. The three methods differ in time, ease of implementation, performance attributes, and expense. Customized programming is, for example, the most time-consuming and expensive approach to the implementation of CAR systems, although it often results in a product that is better suited to the user's requirements. As an alternative, pre-written program packages can be leased or purchased from service bureaus or other companies that specialize in software development. Such pre-written programs are designed, of course, to operate within a specified hardware configuration, and additional constraints may be imposed on data preparation, retrieval, or other user-performed activities.

A further problem generally arises in the identification of potentially suitable software packages. This problem is not limited to CAR applications, but is characteristic of pre-written software in general. The few available software directories are of little help in this respect. Many corporate or institutional computing centers, however, have purchased general data base management or information retrieval programs. Such programs often can be adapted to the requirements of CAR applications.

Turnkey systems provide the fastest and, in some respect, the simplest approach to the implementation of CAR applications. As previously defined, a turnkey system consists of an integrated, pre-selected configuration of hardware and pre-written software. A complete turnkey CAR system includes both computer and micrographics equipment, some of which may be specially designed. In addition to rapid implementation, such turnkey systems offer the advantage of simplified procurement and single vendor responsibility during system installation and subsequent operation. In a customer-developed, multi-vendor CAR system, it may prove difficult to determine the source of, and the maintenance responsibility for, system malfunctions. A given system failure may be attributable, for example, to micrographics equipment, computer equipment, peripheral devices, computer software, or some combination of these components. The disadvantages include the fact that turnkey CAR systems, by definition, provide little or no flexibility in the selection of hardware components, and the customer must sometimes compromise application requirements or modify work performance methods to comform to software constraints.

Performance Characteristics

Whether custom-developed, purchased pre-written, or incorporated in a turnkey system, CAR software typically consists of a series of programs designed to accomplish three broad tasks: (1) capture index data pertinent to specific document images; (2) add to, delete, edit, or otherwise modify previously entered index data; and (3) indicate the microfilm locations of document images pertinent to specified retrieval parameters.

Most CAR systems, as previously noted, are designed for online, real-time operation in which index data entry and subsequent retrieval are performed using terminals. It is possible, however, to implement CAR systems on an offline basis. In such systems, for example, index data may be keypunched or captured by key-to-tape devices for later batch processing and output in the form of a printed keyword or other index. The offline approach is suitable for applications with straightforward retrieval requirements. As an example, the retrieval of microfilmed engineering drawings can be facilitated by a printed Keyword-In-Context (KWIC) or Keyword-Out-of-Context (KWOC) index with entries derived from words used in the title block of each drawing. In such applications, the printed index entries would be accompanied by a microfilm address in the form of an aperture card number or a roll and frame number. The retrieval performance characteristics of such printed CAR indexes are necessarily more limited than those provided in online systems. Most notably, the logical coordination of subjects cannot be accomplished conveniently.

The performance characteristics of online CAR systems are best described with reference to a hypothetical but realistic example. Assume that the customer affairs office of a large company is considering the implementation of a CAR system for the control and retrieval of a product-related file of customer correspondence. Each item of correspondence must be retrievable by various combinations of the following parameters: date, customer name, product name(s), the name of the store where the product was purchased, and the name of the geographic region where the store is located. Customer service representatives who use the file must be able, for example, to retrieve: (1) an item of correspondence written by a specified individual on a given date; (2) those items that reflect complaints associated with a given product in a certain geographic region; or (3) those items that reflect difficulties in a specified store during a specific time period. Other combinations of retrieval parameters are, of course, possible, and it is obvious that such retrieval requirements cannot be handled easily or successfully by a conventional paper-oriented correspondence file.

In the application described above, individual items of correspondence would be recorded using an appropriate source document camera on 16-mm blip-encoded microfilm. For simplified handling at retrieval time, the processed rolls would be inserted into microfilm cartridges or cassettes. Alternatively, microfiche could be used for this application. In either case, it is important to note that correspondence is filmed in random order, thus eliminating the file establishment and maintenance routines typically associated with paper documents. The interrelationships of individual documents are reflected in a machine-readable index, the data for which are entered at an online terminal by an operator working from the documents themselves or, in some systems, from microfilm images of the documents. The CAR software uses the entered data as the basis for establishing and/or updating a series of indexes. In this example, one index would be established for each of the five retrieval parameters noted above: date, customer name, product name, store name, and geographic region. Each index consists of individual entries for index values represented in the microfilmed documents. Thus, the entry of the following index data:

    Date: 11/27/80
    Customer: Appleton W
    Product: Cookware

Store: Cleveland
Region: Midwest

would generate entries in the date index under 11/27/80, in the customer index under Appleton W, in the product index under Cookware, in the Store index under Cleveland, and in the region index under Midwest. Each entry is accompanied by the microfilm and frame address where the indexed document image is recorded. Assuming that the document described in the above indexing example appears on cartridge number 43 at frame number 1382, then the customer name index can be:

Name	Cartridge/Frame
Appleton W	43/1382

In many cases, it is likely that more than one microfilm address will be associated with each entry, reflecting the fact that multiple items of correspondence are indexed with that value. Thus, for example, the date index might be:

Date	Cartridge/Frame
11/27/80	43/1359
	43/1382
	43/1568
	44/0062
	44/0078
11/28/80	44/0347
	44/0349
	44/1087
	45/0003

The above example is necessarily abbreviated, but the same principle applies to the other indexes.

The ability to search an online machine-readable index for the addresses of potentially relevant microimages distinguishes CAR systems from conventional approaches to both document filing and micrographics retrieval. An effective CAR system depends, however, on the development of search software that will interactively guide the user in the successive entry of commands which result in the selection and narrowing of index parameters. The eventual output of such a search is a listing of microfilm or fiche locations. The listing is displayed and/or printed at the user's terminal. Some CAR systems also provide a document description consisting of a listing of index values and, where the documents lend themselves to it, an abstract or other summary information. In some cases, the information in this brief document description satisfies the user's retrieval requirements; in others, it enables the user to weed out obviously irrelevant documents prior to retrieving their microimages.

In terms of specific retrieval capabilities, the typical CAR system will perform the following operations:

1. Search the online index for the film or fiche locations of document images meeting a single specified retrieval parameter. Thus, a command of the form SEARCH CUSTOMER-APPLETON W, will cause the CAR system to search the Customer Name index for the locations of document images which are indexed under APPLETON W. As noted earlier, the index links the various index values to microfilm or fiche locations.

2. Search the online index for the film locations of document images meeting a more complex retrieval parameter involving combinations of several index terms. As noted earlier in this discussion, the ability to perform searches involving the logical coordination of index terms distinguishes online CAR systems from less sophisticated approaches to document retrieval. Thus, a command of the form SEARCH CUSTOMER-APPLETON W *AND* DATE-11/27/80 will cause the CAR system to search the Customer Name index for the locations of document images which are indexed under APPLETON W and the Date index for the locations of document images which are indexed under 11/27/80. The system then will select those document images that are common to both groups. The *AND* operator employed in this example finds the logical intersection of two sets. While the *AND* operation is the most widely used, CAR systems generally permit logical coordination to establish the union of two sets (the *OR* operation) and the negation of two sets (the *NOT* operation). The AND, OR, and NOT operators sometimes are referred to as *Boolean* operators and the process of using them in retrieval operations is often called Boolean logic.

The CAR system's exact response to the search commands described above will necessarily vary somewhat from implementation to implementation. In most cases, however, the initial response is a report of the number of document images satisfying the specified retrieval parameters. This feature, which is called *hit prediction,* allows the user the opportunity to narrow the selection if the retrieved number of items is excessive. When an appropriate number of presumably relevant items is identified, the CAR system will display or print their microfilm addresses, either one at a time or in a list. In some cases, as previously discussed, a brief document description may be displayed as well. Once the microfilm cartridge and frame locations—or microfiche and frame locations—of presumably relevant document images are determined, the operator views and/or prints those images on a reader-printer, using the approaches discussed earlier in this chapter.

5
Reprographics

Office copiers and duplicators • *Phototypesetting*

The preceding chapters have discussed technologies and equipment for the creation, processing, storage, and retrieval of information in an office environment. This and the following chapter discuss, respectively, the reproduction and distribution of information—two activities that are critical to the success of office operations. The focus of this chapter is on reprographics—that field which, broadly defined, is concerned with all types of document reproduction technologies and equipment. On the one hand, the scope of reprographics includes simple document reproduction techniques, such as the use of carbon paper to make copies as a byproduct of typing, or other document creation methodologies. At the other extreme, it extends to the high-speed printers which, as briefly described in the preceding chapter, generate paper copies from machine-readable, computer-processed data. Technically, micrographics can be viewed as a specialized facet of reprographics, as can word processing and certain types of electronic message systems described in the next chapter.[1] In fact, most automated office technologies share attributes that defy rigid categorization by type of information-handling activity.

For purposes of this discussion, however, the scope of reprographics will be limited to technology and equipment for the reproduction of existing paper documents at or near full size. It thus comprises two related product groups of significance in office applications: copier-duplicators and typesetting equipment. These devices are often excluded from discussions of office automation because they produce paper documents. Consequently, it is felt that these devices represent information-processing approaches that are more appropriate to the 1960s than to the 1980s. This extremely limited view of office automation ignores both the important role that reprographic technologies continue to play in office operations—it is virtually impossible, for example, to manage information in an office without a copier or access to one—and important recent changes in these well-established technologies. While, as noted in Chapter One, office automation increasingly relies on magnetic and photographic information carriers, paper documents—and copies of them— are likely to remain important for the foreseeable future.

This chapter reviews reprographic technology, emphasizing those recent equipment developments that are most significant for office automation. It begins with a discussion of copier-duplicator technology and equipment capabilities. Later sections deal with typesetting technology.

OFFICE COPIERS AND DUPLICATORS

Electrostatic Technology

As historically and currently practiced, document reproduction is a photographic activity in which sensitized material and the document to be copied are exposed together to a radiation source, such as light or heat. Although over 20 different copying processes have been more or less successfully used in office applications, the electrostatic process is currently the dominant document reproduction technology. In the electrostatic process, the sensitized material is a photo-receptive surface which is capable of accepting and retaining an electrical charge in darkness, as well as selectively releasing the charge on exposure to light reflected from the document to be copied. The reflected light typically is directed to the photoreceptive surface through a lens. Dark (text) areas of the document absorb light while light (background) areas reflect it. Electrical charges consequently are dissipated in areas of the photoreceptor that correspond to nontext areas of the document. The remaining charges form a latent image of the document's text. This latent image is made visible through the application of oppositely charged, finely ground carbon particles called *toner*, which may be in either powdered form or suspended in a liquid.[2]

Within the broad outline delineated above, the electrostatic process has two variants. In the most popular—the indirect, or *xerographic*, variant—the latent image first is created and developed on an intermediate surface, such as a selenium-coated drum. The charged toner particles that form the image then are transferred to an oppositely charged sheet of ordinary paper to which they are fused with heat or pressure. Because the resulting copy is on uncoated paper, the xerographic process enjoys high user acceptance. Originally developed by the Xerox Corporation—which, through the 1960s, was the sole American manufacturer of plain paper copiers—the xerographic process now is employed in the equipment of more than a dozen domestic and foreign copier-duplicator manufacturers.

As originally developed by RCA in the late 1950s, the direct electrostatic or *electrofax* process employs a photoreceptive sheet of paper coated with zinc oxide on which a latent image is created in the manner described above. Toner then is applied and fused to the photoreceptor which itself becomes the copy. Although photoreceptive paper is of a higher quality and consequently more expensive than uncoated paper, electrofax equipment is much plainer and less expensive than its xerographic counterpart. Through the mid-1970s, the total cost of electrofax equipment rental or purchase and consumable supplies generally proved to be much lower than that of more complex xerographic copiers in the low to medium volumes (1,000 to 10,000 copies per month) characteristic of many office applications. Thus, despite a strong user preference for xerographic copies, electrofax equipment was and is widely used in offices. Recently, however, several American and Japanese manufacturers have introduced comparatively simple, lower cost xerographic copiers which compete more favorably with electrofax equipment in low- to medium-volume applications. Increased utilization of these devices has led to a decline in office placements of electrofax equipment.

Equipment Design and Operation

The electrostatic process, or modified versions of it, is used in several types of automated office systems, including those described in preceding chapters. A number of available micrographic reader-printers, for example, employ electrostatic technology. Similarly, the A.B. Dick/System 200 Record Processor, an updatable microfiche system described in Chapter Four, employs a variant of the electrostatic process to create reduced-size document images on a transparent film base. Likewise, the Xerox 9700 and IBM 3800 page printers use the xerographic process in conjunction with lasers to produce paper reports at high speed from machine-readable, computer-processed data. As described in Chapter Six, laser imaging also is used in those facsimile transceivers that employ xerographic reproduction.

The focus of this section is, however, electrostatic equipment which makes copies of paper documents. The reprographics industry divides such equipment into two broad groups: *copiers,* which are designed to make one or several reproductions of a document; and *duplicators,* which are designed to make many copies of a document. The distinction is necessarily imprecise. The point at which a duplicator is preferable to a copier is variously set at 10, 15, or more copies per original, and the distinction is often ignored in actual practice. Copiers and duplicators, however, do differ in their multiple-copy capabilities. The typical electrostatic duplicator delivers the second and subsequent copies of a given document at a much faster rate than a copier will. Duplicators thus enjoy significant productivity advantages in applications involving numerous copies of single or multiple documents. As might be expected, duplicators are typically larger, more complex, and more expensive than copiers. Since the two devices deliver the first copy of a document at or near the same speed, the additional expense associated with a duplicator is only justifiable where the faster multiple-copy capabilities

Figure 5-1. Copiers are machines designed for the convenient single- or multiple-copy reproduction of documents. The Minolta Electrographic 301 is one of a new generation of desk-top electrostatic copiers which substitute fiber optics for the lens and light source used in conventional copying equipment. (Courtesy: Minolta Corporation)

will be utilized frequently. Duplicator rental and lease pricing plans typically are designed to encourage long-run, high-volume reproduction,

The typical duplicator is a free-standing, immobile device which occupies a relatively large amount of office floor space and requires an electrical outlet, air conditioning, and other special site preparation. Electrostatic copiers, by way of contrast, are available in both floor-standing and table models. Most are designed to operate from an ordinary electrical outlet in an office environment, although power fluctuations or extremes of temperature and humidity may affect quality and/or reliability adversely. There is a continuing trend toward quieter copier operation and more compact equipment design. In several newer machines, the conventional lens, mirrors, and related optical components are replaced by a simple fiber optics assembly in which a cluster of microscopic glass strands directs reflected light from the document to the photoreceptor. Measuring about the size of an office typewriter, the machines are energy-efficient and extremely compact—an important consideration in office applications. Some copier manufacturers, recognizing the office user's need to respond to changing application requirements, mount their equipment on wheeled pedestals or stands to facilitate mobility.

Apart from differences in size and multiple-copy speed, electrostatic copiers and duplicators are similar in operation. The document to be reproduced is laid face-down, either manually or automatically, on a glass

Figure 5-2. While copiers are designed to make one or several reproductions of a given document, duplicators are intended for applications requiring a number of copies of each document. As a result, duplicators are typically much faster than copiers, especially when making the second and subsequent reproductions of each original. The Xerox 8200 is an electrostatic copier/duplicator designed for office operation. (Courtesy: Xerox Corporation)

imaging platen. The operator uses control dials or buttons to select the number of copies to be made, then to activate the exposure lamp. With most equipment, the platen remains stationary while a light source travels the length of the document, exposing it in its entirety. Alternatively, the platen itself may move the document past a stationary light source. Such traversing copiers, which are used widely in inexpensive copiers, are well suited to the reproduction of cut sheets of paper, but cannot readily accommodate bound volumes. Regardless of the platen type, light from the document is directed to a photoreceptive surface and a latent image then is developed in the manner described earlier in this chapter. The resulting copy is delivered and stacked in a tray typically located in the front or at the side of the machine.

Compared to other types of office equipment, copiers and duplicators have proven vulnerable to a surprisingly wide variety of major and minor malfunctions, most of which limit the machine's operation and necessitate a service call. Lacking a self-contained photoreceptor and associated components, electrofax copiers are generally more reliable than their xerographic counterparts. The reliability of plain paper copiers and duplicators has recently improved, however. The tendency toward more compact design has shortened the distance the copy paper must travel during image development and fusing, thus minimizing the potential for jamming—a frequent source of downtime in older equipment. When paper jams do occur, they often can be cleared easily by the operator.

While electrostatic copiers and duplicators remain essentially mechanical devices, manufacturers are increasingly replacing conventional dials and other controls with reliable electronic components. In the manner of the microfilm cameras described in Chapter Four, a growing number of copiers and duplicators incorporate microprocessors as controllers which monitor machine operations, control the distribution of toner, diagnose the cause of malfunctions, and direct the operator in the performance of simple repairs, thus eliminating needless service calls. With microprocessor control and a small memory, a multiple-copy job interrupted by a paper jam or other malfunction can be resumed at the point of interruption when machine operation is restored. In some copiers, the electronic memory also allows an operator to interrupt a long job conveniently to perform one or several shorter ones.

Reprographic Capabilities

While the electrostatic process dates from the 1950s, the mid-1970s was a time of important developments in both electrostatic technology and equipment design which resulted in the significantly improved reprographic capabilities that copier and duplicator users now take for granted.[3] Since their introduction, electrostatic copiers have been designed to reproduce the letter (8.5- by 11-inch) and legal (8.5- by 14-inch) documents prevalent in business applications. Some models are loaded with rolls of paper from which individual copies automatically are cut to the size of the original. Roll capacities in excess of a thousand letter-size copies are common with high-speed duplicators. Sheet-fed copiers and duplicators, as the name implies, are loaded with stacks of pre-cut paper in either letter or legal size. The capacity of the machine's paper tray varies from several hundred to several thousand sheets. While sheet-fed equipment often can copy onto office letterheads or other pre-printed paper stock, older models required that the operator change the paper supply to accommodate letter or legal copying requirements. Increasingly, however, sheet-fed models are equipped with dual trays or cassettes which enable the copier to contain multiple paper sizes and stocks simultaneously. The appropriate paper supply is selected by a button or dial.

An increasing number of copiers and duplicators can reproduce documents larger than the customary maximum of 8.5 by 14 inches. For the most part, these large document copiers were developed in response to the increasing prevalence of 11- by 14-inch computer printouts in office applications. A few machines, designed primarily to meet the requirements of accounting offices, will reproduce ledger-size (14- by 18-inch) or even larger documents. As an alternative to size-for-size reproduction, several manufacturers offer reduction copiers and duplicators that reproduce large documents onto letter-size paper. Unlike microform reproductions, the resulting copies are approximately two-thirds to one-half the size of the original and are legible with the unaided eye. Most available reduction copiers will accommodate documents up to 14 by 18 inches in size. Alternatively, two letter-size documents can be copied, side by side, onto a single sheet of letter-size paper. The most complex and expensive machines offer a choice of several reductions. Platen markings and color-coded instructions guide the operator in selecting the reductions necessary to produce a letter-size copy from documents of various sizes. Simpler, less expensive devices permit only one reduction, typically sufficient to reduce a computer printout to letter-size. Most reduction copiers and duplicators also will make size-for-size reproductions, although a few machines invariably copy in the reduction mode.

Because they save filing space, mailing weight, and, in some cases, reproduction costs, reduction copiers and duplicators have enjoyed considerable recent popularity in office applications. Similar effects often can be obtained with two-sided copying. Several available plain paper copiers automatically will copy onto both sides of a sheet of paper. Electrofax copiers have been limited traditionally to single-sided copying because the photoreceptive coating essential to image formation typically appears on only one side of the copy paper.

With regard to improvements in copy quality resulting from refinements in electrostatic technology, most newer copiers and duplicators can make satisfactory reproductions of half-tone or continuous-tone illustrations as well as textual documents. Likewise, developments in exposure control and toner distribution systems permit the making of consistently useable copies from a wide range of office documents. Most new xerographic copiers and duplicators have a density control dial or button—long a standard feature with electrofax machines—to allow the operator to compensate for variations in the contrast of documents to be copied. As previously noted, exposure control and toner distribution are increasingly regulated by microprocessors.

Xerographic machines, once notorious for the inability to reproduce certain shades of blue, now routinely respond to all colors, reproducing them in black and white. Where the ability to make color copies is required, the Xerox 6500 combines magenta, cyan, and yellow toners to create several colors as well as black. The 6500 also can be equipped with an adapter for making paper enlargements from 35-mm color slides. At present, relatively little office information is in color. However, with the advent of color display terminals, the need for color document reproduction capability may increase.

Productivity Features

In keeping with the emphasis on the automation of previous manual office tasks that characterize other types of office equipment discussed in this book, newer electrostatic copiers and duplicators are increasingly incorporating standard or optional features designed to improve operator productivity. When a copier is equipped with a document feeder, for example, the operator is freed of the tedious time-consuming task of positioning the individual sheets of multi-page documents on the copier's glass platen. The typical document feeder consists of a set of rubber rollers or belts designed to transport manually inserted documents past an imaging area. By inserting originals continuously into the feeder, an operator can achieve impressively high work throughput rates, even with machines which, when operating in the conventional platen-mode, might be considered slow. Some document feeders incorporate a stacker designed for the automatic insertion of specified numbers of documents. While the document feeder and related accessories are sometimes fixed in position, most machines allow the operator to lift or otherwise move the feeder out of the way to copy bound materials or large documents. There is likewise a tendency toward modularity in machine

Figure 5-3. Automated accessories, designed to increase operator productivity, are an increasingly common feature of electrostatic copiers and duplicators. The Kodak Ektaprint copier/duplicator features an image positioner which uses a vacuum-belt feed mechanism to automatically position documents on an imaging surface and return them to the operator after a specified number of copies have been made. (Courtesy: Eastman Kodak Company)

design, which permits the customer to order the copier first and later add a document feeder as application requirements dictate.

Document feeders, as noted above, are particularly useful when a single copy is to be made of a multi-page document, although several feeders are compatible with multiple copying of many documents. Whether a feeder is used or not, several standard or optional accessories are designed to facilitate the production of finished sets when making multiple copies of a number of originals. A sorter, the earliest and still the most important of these accessories, is designed to eliminate the manual labor associated with placing individual copies of multi-page documents in sequence to create separate sets for distribution. When a copier or duplicator is equipped with an integral or attached sorter, multiple copies of successive pages are routed automatically to separate compartments or bins where they are stacked in sequence, awaiting removal by the operator. The number of available bins varies from 10 to 50 or more and is expandable. Once available only on high-speed duplicators, sorters now are offered as optional accessories on

Figure 5-4. To improve operator productivity, office copiers, like the Sharp SF-850 here, are now equipped with sorters, document feeders, and other features previously associated with high-volume duplicating equipment. (Courtesy: Sharp Electronics Corporation)

many medium-volume copiers. As an alternative to sorting, some copiers and duplicators feature a recirculating feeder that recopies multi-page documents in sequence to create the required number of sets.

Several electrostatic duplicators automatically will staple sorted copies, thus producing completely finished sets.

Non-Electrostatic Equipment

While the electrostatic process, as previously noted, dominates document reproduction activity in office applications, other types of reprographic equipment remain available. These alternative technologies are typically less automatic than electrostatic equipment and require greater operator skill and intervention. In *offset duplication,* the dominant technology in "print shop" environments, the operator must first prepare a master by direct impression (typing or handwriting) or by using a photographic or electrostatic mastermaker. Paper masters generally can be used for runs of 200 copies or less. These masters can be created with an electrostatic copier. Aluminum plates are required to maintain copy quality in longer runs. Once created, the master is attached to the top cylinder of an offset press where it is inked and used to print an image onto a second cylinder. This second cylinder then is used to print the copy onto a sheet of ordinary paper.

While early offset machines required considerable operator attention to such tasks as master handling and ink adjustments, the newer duplicators are highly automated. Many of them feature microprocessor/controllers that perform certain diagnostic and quality control procedures. While large, high-speed offset presses typically are located in centralized reproduction facilities, smaller desk-top units are available for office installations. These compact units will produce several thousand high-quality copies per hour.

While offset technology is noted for very high copy quality and low unit costs in long runs, the formerly dominant position of offset equipment in high-speed duplication is increasingly challenged by electrostatic equipment. The Xerox 9400, for example, is a fully automated electrostatic duplicator that is designed for use in high-volume applications previously employing offset presses. The 9400 is capable of automatically feeding several hundred originals of varying sizes and thicknesses, copying onto one or both sides of a sheet of paper at speeds up to two pages per second, and sorting an infinite number of sets. Unlike offset equipment, the 9400 does not require the making of a separate master or the use of a technician as an operator.

Other document reproduction processes retain a very limited role in office applications. In *spirit duplication,* for example, a paper master backed with an image formed from a carbon dye is moistened with a solution of methyl alcohol to print multiple copies. In *mimeography,* a stencil-like master is created by typing or otherwise removing the ink-impervious coating from

Figure 5-5. While electrostatic copiers and duplicators attract the bulk of users' attention, off-set duplication remains an important reprographic technology. Newer offset presses are highly automated and incorporate microprocessors and related integrated circuitry to improve flexibility and reliability. The AM TCS/System 4 can produce up to 9,000 copies per hour. (Courtesy: AM Multigraphics)

a fibrous-based tissue. When mounted on a cylindrical inking pad, ink is forced through the stencil onto paper. While newer spirit duplicators and mimeograph machines are microprocessor-controlled, these two technologies are increasingly displaced by electrostatic equipment in office applications involving short-run duplication.

Among copying processes, the once popular heat-based technologies—notably, the dual spectrum process and thermography—rarely are used in office applications, having been displaced almost completely by low-volume electrostatic copiers. Heat-based technologies, such as the vesicular and thermal silver processes, remain important, however, in micrographics applications. The diazo process continues to be used in engineering office applications requiring the reproduction of drawings and other large, translucent documents. Despite disadvantages associated with the use of ammonia, the diazo process remains the only means of readily duplicating D-size (24- by 36-inch) or larger drawings. While electrostatic copiers and duplicators, as noted in an earlier section, are increasingly capable of accepting large documents, the maximum acceptable size is typically 14 by 18 inches. As is true of the heat-sensitive processes mentioned above, the diazo process is used widely for the reproduction of microforms. As discussed in Chapter Four, several desk-top diazo microfiche duplicators are available for office applications.

Intelligent Copiers

Copiers and duplicators often are viewed as "mature" office products, their performance characteristics having developed to the point where equipment improvements merely constitute refinements of existing functions. As an example, the incorporation of microprocessors as controllers in most newer machines has improved reliability and simplified certain operating procedures, but has not added significant new capabilities. In some respects, electrostatic technology leaves little scope for further development. It is unlikely, for example, that copy quality can be improved significantly; it is already excellent.

Yet, considerable recent interest, and some confusion, has been generated by a new type of reprographic device that is loosely called an "intelligent copier."[4] This emerging product actually consists of two types of related equipment:

1. *Intelligent printers*. These are microprocessor-controlled devices that employ electrostatic technology to generate paper documents from machine-readable data prepared by computers or, in some cases, by automated text editing systems; and

2. *Intelligent copier-printers*. These copiers are likewise microprocessor-controlled and employ electrostatic technology, but can function as both an office copier and a printer linked to computers or automated text editing systems.

The performance characteristics of intelligent printers—sometimes called page printers—were introduced briefly in Chapter Four where their relationship to COM recorders was discussed. They are designed primarily as replacements for conventional line printers in centralized computer installations. The Xerox 9700, the most sophisticated example of this type of product, accepts machine-readable data and uses a laser to create latent images of individual characters out of a pattern of electrostatic charges on an intermediate surface. In the manner of conventional xerographic copier-duplicators, the latent image is developed through the application of toner, and the resulting copy then is transferred to a sheet of plain paper. To produce copies of sufficiently high quality, individual characters are created out of an extremely dense dot matrix. A very wide range of type fonts and sizes are possible, and forms also can be generated directly from digital input. The Xerox 9700 can operate online to certain IBM computers and offline, via a magnetic tape interface, with computers of other manufacturers. It also can accept input, via telecommunications, from a Xerox 850 text editing system. As noted in Chapter Four, a microfiche subsystem is optionally available. Laser printing also is used in the IBM 3800, another xerographic page printer designed for high-volume computer output applications. The Wang Intelligent Image Printer (IIP), by way of contrast, uses the xerographic process in conjunction with a fiber optics character generator.

It is designed for use as an output device in the Wang Office Information System.

Introduced in 1979, the IBM 6670 Information Distributor is an example of an intelligent copier-printer. Similar to the Xerox 9700 and IBM 3800, it uses a combination of xerographic and laser technologies to generate plain paper copies from machine-readable, computer-processed data received via telecommunications at speeds up to 4800 bits per second. It also can operate online to another IBM 6670 or to an appropriately equipped IBM magnetic card typewriter or Office System 6 word processor. The IBM 6670 can generate high-quality characters in four different type-styles and several horizontal spacings. It is capable of two-sided copying and can support a magnetic card reader with a stacker capable of accepting up to 50 cards. The 6670 is designed for office installations and is further distinguished from intelligent printers by its ability to function as a convenience copier.

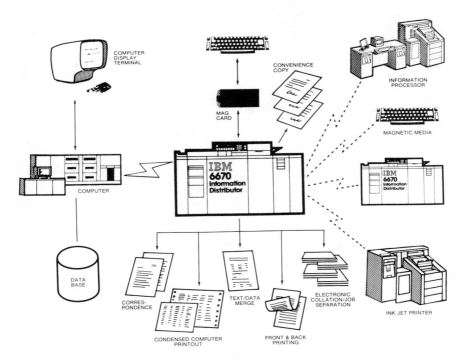

Figure 5-6. Intelligent copiers represent a merging of reprographic and computer technologies and are likely to be an important integrating device in the office of the future. The IBM 6670 can reproduce documents in the manner of conventional copiers. It also can communicate with electronic devices—computer systems, word processing systems, various printers, or other 6670s. (Courtesy: IBM Office Products Division)

In that mode, it operates at speeds up to 36 copies per minute. Being microprocessor-controlled, it can interrupt the printing of machine-readable data or text to permit copying of documents, then resume the interrupted function when copying is completed.

The Xerox 5700 is, like the IBM 6670, an intelligent copier-printer designed for office applications. As a result of the combination of xerographic and laser recording utilized in the Xerox 9700, it can operate as a copier and accept machine-readable input from computers, text editing systems, or other intelligent copier-printers. It can generate forms, including graphics

Figure 5-7. Intelligent copiers can perform many functions traditionally associated with phototypesetting equipment. The Xerox 5700, for example, offers the operator a choice of type fonts and sizes. It can communicate with computer and word processing systems as well as reproduce paper documents. (Courtesy: Xerox Corporation)

and company logos, in addition to high-quality textual information. As a copier, it is capable of two-sided reproduction, sorting, and stapling. A unique feature of the 5700 is a touch-sensitive display screen that replaces conventional buttons and other controls as the method by which the operator initiates various machine functions.

While the two machines described above utilize lasers, several Japanese manufacturers—including Minolta, Canon, and Konishiroku—are developing intelligent copier-printers which employ the xerographic technology in combination with fiber optics. These devices, which are expected to enter the American market in the early 1980s, will be designed for office installations using text editing systems and/or small business computers. They typically will operate at speeds in the 10 to 20 pages per minute range. In addition to functioning as convenience copiers and digital printers, these devices will be able to create new documents by merging information from existing paper documents with machine-readable data prepared by computers or text-editing systems.

PHOTOTYPESETTING

Typesetting Technology

The intelligent copiers and computer page printers described in the preceding section often employ multiple type styles and sizes to enhance the appearance and, in some cases, the utility of office-generated paper documents. As such, these copiers and printers offer document creation capabilities normally associated with typesetting equipment. While documents produced by typewriters, word processing equipment, and computer-driven printers are satisfactory for most office applications, such office-generated publications as annual reports, catalogs, and newsletters require the aesthetic and functional properties associated with typeset text. A typeset document is undeniably more impressive than a typewritten one. Even when proportional spacing and interchangeable printing elements are used, typewritten pages cannot match the attractiveness and variety of typeset text. More importantly, typesetting often improves the utility of a document by facilitating reading and emphasizing significant points. The use of bold headings, for example, can make the organization of a report more readily comprehensible, while italicized words or phrases can call the reader's attention to significant concepts. Similarly, in catalogs and printed listings, varied typographic capabilities can distinguish primary and secondary data elements, thus facilitating the search for desired information.

Although it remained comparatively static for the 400 years following Gutenberg's innovative work with movable type, typesetting technology has changed considerably since 1900.[5] The then prevalent method of setting lines of "hot" type by pouring molten lead into metal molds has been displaced largely by alternative technologies. In *direct impression* typesetting, often described as "cold" typesetting, a typewriter-like impact device, such as one of the several models in the IBM Selectric Composer or AM Varityper Composer product lines, prints individual characters as they are entered by an operator at an attached keyboard. The characters can be recorded on paper or direct impression offset masters. These cold type devices are more convenient and less expensive, in terms of both equipment and labor, than the setting of hot type. The quality of the resulting output, while somewhat lower than that obtainable with hot typesetting, is adequate for most applications. As an additional advantage pertinent to this discussion, most impact composers are suitable for desk-top operation in an office environment.

Impact composers are limited, however, in several potentially significant respects: (1) the maximum type size is typically 12 points, making them unsuitable for headline work; (2) while the printing elements may be interchangeable, only one type style can be online at a given time; (3) error correction and character fitting can prove difficult; and (4) proper justification may require double typing.

While some of the limitations of earlier electromechanical equipment have been minimized or eliminated in the newer electronic models, office applications are increasingly utilizing phototypesetters in preference to impact composers. While often described as an alternative cold type methodology, phototypesetting is actually a non-impact typesetting technology. Unlike impact composers with their metal type faces, the typical phototypesetter maintains a master matrix of type fonts on film disks, drums, strips, or similar photographic media. In response to a combination of text and commands entered at an attached keyboard or from externally prepared machine-readable media, the appropriate characters from the master matrix are projected, in a specified sequence and position, onto a photographic film or paper master. Available machines vary in the number of type fonts they maintain online and in the range of sizes at which a given font can be projected. The least expensive phototypesetters are comparatively simple, manually operated lettering machines designed to generate headline or title characters for newsletters, report covers, promotional materials, and similar office-generated documents. As is true of more sophisticated equipment, the least expensive phototypesetters set type by projecting light through internally stored character images onto photosensitive strips or sheets of film or paper. Their online typographic repertoire, however, typically is limited to one or two fonts and styles, although reproducible character images generally are represented on interchangeable matrices.

More complex phototypesetters are designed to generate text as well as headlines. Depending on the machine, input may be directly entered at an attached keyboard. Alternatively, a combination of machine-readable commands and characters can be prepared on separate devices and encoded on magnetic or paper tape. The encoded commands direct the machine in the automatic selection, positioning, projection, and recording of individual characters on photosensitive paper or film. Multiple type fonts often can be maintained online, and the size of projected characters can be controlled within certain limits. Similarly, the project method can be modified to achieve such special effects as the conversion of roman to italic type and the stretching of lines to make them appear longer or higher.

As an alternative to electromechanical phototypesetters that use lenses and mirrors to project characters from internally stored photographic matrices, the "electronic" phototypesetters generate individual characters and type fonts from digitally stored definitions. They demonstrate the impact of recent developments in computer and reprographic technologies on the typesetting activity. Operating under the control of a dedicated

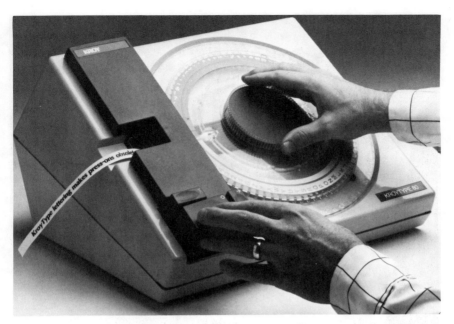

Figure 5-8. The Kroy 80™ is a direct-impression typesetter that is well suited to office applications involving the labeling of reports, files, slides, or microfilm containers. It also can be used for the titling of microfiche and other flat microforms. It produces instant "type on tape" in sizes from 8 to 36 points. (Courtesy: Kroy Industries, Inc.)

minicomputer or microcomputer, the electronic phototypesetter displays text on an integral cathode-ray-tube screen which, in turn, is used to expose a photosensitive paper or film. One such device—the COMp80/2 developed by Information International Incorporated—records typeset text on microfilm or fiche.

Regardless of the photosensitive medium, the characters themselves are formed out of predefined patterns of closely spaced lines. The digitized storage of character definitions and the resulting elimination of photographic matrices permit significantly faster operation. These electronic phototypesetters likewise offer a greater range of online font generation capabilities and simplified incorporation of additional fonts. As might be expected, however, electronic phototypesetters are significantly more expensive than their electromechanical counterparts and their use is limited to the high-volume applications encountered in the printing and publishing industries. Much of the higher equipment cost is attributable to the complicated process of developing digitized font definitions and the relatively large amount of online storage space required for those definitions.

A variant form of electronic phototypesetter replaces the CRT screen

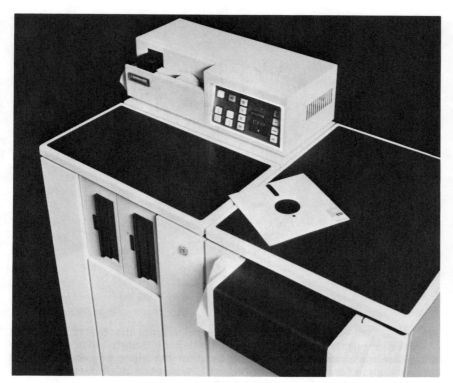

Figure 5-9. Responding to combinations of text and commands recorded on externally-prepared machine-readable media, phototypesetters record characters on photo-sensitive film or paper masters. The earliest generations of phototypesetters projected character images from internally-stored photographic matrices. The Merganthaler Linotron 202, however, generates individual characters and type fonts from digitally-stored definitions. Operating under the control of a dedicated computer, it displays text on an integral CRT screen which is, in turn, used to expose a photosensitive paper or film. This book was typeset with a Linotron 202. (Courtesy: Merganthaler Linotype Company)

with a laser beam that "images" the photosensitive medium directly. These devices offer the theoretical advantage of being able to generate half-tones by scanning continuous-tone copy, although that capability has not been implemented on equipment that is presently available. Similar to the CRT-based phototypesetters, these laser-oriented devices offer the greater reliability that is inherent in equipment with relatively few moving mechanical components. As is true of the laser beam COM recorders and page printers discussed in previous sections of this book, individual characters are generated out of a matrix of closely spaced dots from digitally stored definitions.

Phototypesetting and Word Processing

In most offices, typesetting is a periodic or sporadic requirement which is met by submitting typewritten text to a centralized inhouse reprographic department or to a typesetting service company. In doing so, the office often must sacrifice control over work scheduling and turnaround. The time required for the delivery and pickup of work further extends the lead time required for a given job, while the associated inconvenience may lead to underutilization in applications where typesetting would improve the appearance and utility of office-generated information. In addition, the keyboarding of text, performed once in the originating office, generally is repeated by the typesetter operator, thus increasing the time and cost required to complete the work. Many of these problems could be resolved if some or all of the typesetting work were performed in the originating office, but with prices for high-speed electromechanical or electronical phototypesetters routinely exceeding $50,000, inhouse typesetting only has proven justifiable in very high-volume applications. In addition, many phototypesetters require special operator skills not possessed by office clerical personnel.

Recent technological developments, however, warrant a reconsideration of the practicality of inhouse phototypesetting. These developments, which are described in the remainder of this section, are of two types: (1) several manufacturers have developed relatively low-cost, direct-entry phototypesetters suitable for office use; and (2) an increasing number of phototypesetters can accept input prepared by automated text-editing systems, thus eliminating the need to rekeyboard typed text prepared by the originating office.[6]

As noted earlier in this chapter, depending on the particular model, available electromechanical and electronic phototypesetters will accept input, in the form of commands and characters, entered at an attached keyboard or recorded on magnetic media or punched paper tape by an external device. Those devices that accept input from an attached keyboard are called *direct-entry* typesetters. In the manner of impact composers, they function with a typewriter-like keyboard. The newest models closely resemble the display-oriented text-editing systems discussed in Chapter Two. They operate under microprocessor control, feature a CRT for the display and verification of input, and capture keystrokes on a floppy disk, tape cassette, or other magnetic media for later redisplay and modification. They are preprogrammed with error correction, document revision, and related text-editing and formatting capabilities. Many of them include automatic hyphenation capabilities. Unlike conventional text-editing systems, however, these direct-entry phototypesetters produce typeset rather than typewritten output. Some models will maintain as many as 16 fonts online and generate characters in sizes from 5.5 to 74 points. As is true of display-oriented text-editing systems, the operator can enter work while output is being generated, and some models operate at output speeds sufficient to support multiple input

Figure 5-10. Many newer phototypesetters will support direct entry or auxiliary input stations at which the text to be typeset and related machine instructions are converted to machine-readable form. Such input stations utilize CRT technology and resemble word processors in their text preparation and editing capabilities. The AM Varityper Comp/Set 4510 (right) is a direct-entry phototypesetter capable of generating characters in 16 styles and 70 sizes. The Comp/Set 5404 (left) is a stand-alone input terminal designed for the off-line preparation of text for Comp/Set phototypesetters. (Courtesy: AM Varityper)

stations. The newest models are well suited to office operations and, with prices for a single keyboard system as low as $15,000, they constitute a potentially cost-effective alternative to external typesetting services in medium-volume office applications.

Where external phototypesetting services are used, the interfacing of typesetting and word processing equipment significantly reduces or eliminates the time and expense associated with the rekeyboarding of text. Text typed by the originating office on word processing equipment theoretically should not have to be retyped for input to the phototypesetter. In actual practice, there are two interrelated facets to a word processing/typesetting interface:

1. Machine-readable output from a word processor must be conveyed physically to a phototypesetter; and

2. The word processor output must be translated into a form compatible with typesetter input.

As briefly noted in the preceding section, many phototypesetters will accept input, consisting of a combination of coded commands and characters, on a punched paper tape that is commonly called a teletypesetter (TTS) tape. Tape can be prepared at an offline input station with a specially designed keyboard. Many of those devices feature CRT screens and preprogrammed text-editing capabilities similar to those described in the preceding discussion of direct-entry phototypesetters. With regard to applications involving computerized preparation of text for input to phototypesetters, the TTS tape can be computer-generated as well. Several word processing systems can be optionally equipped with a TTS tape punch which converts the output text to a physical medium for phototypesetter input. Alternatively, some phototypesetters can be equipped with floppy disk drives or tape cassette readers that enable them physically to accept magnetic media containing word processor output. In a few cases, the phototypesetter can accept word processing output directly. More often, an intermediate device serves as a translator, taking the medium recorded on a word processor and converting it to a machine-readable recording suitable for use by a phototypesetter.

As an alternative to the physical conveying of word processor output to the phototypesetter, the word processing system can be equipped with telecommunications capability, as discussed in Chapter Six. Prepared text then can be transmitted electronically to a phototypesetter or to an intermediary device, such as a computer, which will record the transmitted characters on punched paper tape or other machine-readable media acceptable to a phototypesetter. The use of telecommunications in this manner overcomes problems associated with the incompatibility of physical media. As yet another means of physically conveying word processor output to a phototypesetter, text generated by a word processor can be printed on paper in an OCR-compatible type font. The resulting paper documents then can be scanned by an OCR reader which will convert the recorded characters and coded commands to a punched paper tape or other media for input to a phototypesetter.

Regardless of the method by which it physically is conveyed to a typesetter, the output of a word processor consists of a machine-readable stream of characters with embedded machine commands. The commands, as previously discussed in Chapter Two, are represented by predefined codes that instruct the word processor to center the text, align decimals, underline specified words, maintain meaningful hyphens, and perform similar operations. In most cases, the commands that initiate a specific action with a given word processor are different from the commands required to initiate the same action on a given phototypesetter. Thus, a right justification command embedded in word processor output will not necessarily initiate right justification by a phototypesetter. In addition, phototypesetters can generate characters and respond to commands for which there are no word processing counterparts. Some phototypesetter fonts, for example, include ligatures—combined characters, such as a double "f," linked by a common stroke or bar—that are not represented on the typewriter-style keyboard used in word processing. Likewise, open and close quotation marks are represented by the identical character in typewriting, but by two different characters in most typeset fonts. In terms of command codes, there is no word processing equivalent for those commands that instruct a phototypesetter to change type fonts and styles or to set specified characters in italics.

The resulting problems can be overcome in several ways. Word processing input can be prepared, for example, with embedded substitute codes designed to initiate specific typesetting operations. These substitute codes then are translated into their phototypesetter equivalents by an interface device. Several companies specialize in the development of such translator/interfaces which convert command codes generated by specified word processing systems into the command codes required to initiate a given action on a specified phototypesetter. Alternatively, the word processing system can be used for the preparation of text only. Command codes can be added to the text by the phototypesetter operator working at either an offline or direct-entry input station. In this approach, the word processing operator can record text at maximum speed, freed of the necessity of embedding complex typographic code sequences. The phototypesetter operator, who is presumably more familiar with the required commands, simply can embed them in the appropriate places without rekeyboarding the entire text.

Photocomposition

The phototypesetters described in the preceding section produce typeset text on photosensitive paper or film. The text itself typically consists of continuous lines and paragraphs recorded on long sheets without regard to page formats. With regard to conventional phototypesetter output, page layouts are manually created by cutting and pasting the typeset text—an activity called page *make-up*. A *photocomposer* is a more sophisticated phototypesetting device that performs page layouts automatically. As with phototypesetters, the design and operation of photocomposers have been influenced strongly by recent developments in computer technology. As is true of phototypesetting, photocomposition is increasingly performed at a specially designed make-up or composition terminal equipped with a CRT screen. In fact, the interchangeable use of the terms "photocomposer" and "videocomposer" reflect the critical role of the display screen in the page composition activity. The composition terminal can operate either online to a phototypesetter or offline. In the latter case, it generates a machine-readable recording on a teletypesetter tape or a suitable magnetic medium.

The advantages of a videocomposer are obvious: operating under microprocessor control, it enables the operator to visualize, modify, and, in some cases, extensively manipulate page layouts prior to actually setting type. As is true of the newer phototypesetting input stations discussed earlier, videocomposers typically are preprogrammed with text-editing and formatting capabilities, including automatic hyphenation and margin justification. They often provide floppy disks for online text storage and may

149

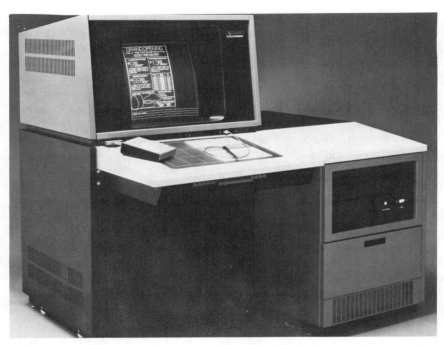

Figure 5-11. A photocomposer is designed to eliminate manual page make-up by performing page layouts automatically with the aid of a CRT display. Such "videocomposers" allow the operator to visualize, modify and—in some cases—extensively manipulate page layouts. The Compugraphic AdVantage II is an extremely powerful and versatile make-up system which uses an electronic input tablet and pen to permit operator manipulation of page images. (Courtesy: Compugraphic Corporation)

accept input from word processing systems, OCR readers, or via telecommunications from compatible devices. In addition to a keyboard, an increasing number of systems provide light pens and electronic graphics tablets and drawing boards which allow the operator to generate graphics, logos, or special form layouts easily. While, at the time this chapter was written, systems suitable for newspaper layout work could cost upward of $100,000, several systems priced in the $20,000 to $30,000 range can meet the photocomposition requirements of most office applications. It is expected that an increasing number of lower cost, office-oriented videocomposers will become available during the coming decade.

CAMIS Technology

The phrase Computer-Assisted Make-Up and Imaging Systems (CAMIS) is used to denote the convergence of reprographic and computer technologies to permit new approaches to the dissemination of information in paper form.[7] Various aspects of this convergence have been suggested in preceding sections of this chapter. Newer electrostatic copiers and duplicators, for example, are increasingly microcomputer-controlled and feature random-access memories for the storage of limited amounts of job-related information. "Intelligent" copiers can generate multiple copies from machine-readable input as well as from paper originals. Operating under microprocessor control, they can interface with computers, word processors, and telecommunications facilities. In their ability to generate multiple-type fonts and sizes, intelligent copiers perform in a manner similar to phototypesetters. Phototypesetters and photocomposers likewise reflect the impact of computer technology. They routinely feature microprocessor controllers, input stations with CRT displays, magnetic media for online text storage, and preprogrammed text-editing capabilities. The direct-entry phototypesetters, discussed in a preceding section, can be viewed as special-purpose word processors designed to produce typeset rather than typewritten text.

Proponents of CAMIS, however, foresee a more comprehensive integration of technologies that will provide an office-based user with greater power and flexibility in the creation and dissemination of information in the form of single or multiple-copy paper documents. In a futuristic scenario for a fully configured CAMIS operation, an office-based worker, equipped with a display terminal operating online to a computer, will be able to: (1) key-enter and edit text; (2) store text on magnetic media, pending later revisions; (3) route text electronically to other terminals for additional input or modification; and (4) select from a range of available typewritten and typeset output formats and devices. In addition, the operator will be able to access data bases or other existing accumulations of machine-readable text for purposes of integrating their content into newly created documents. The display terminals will have attributes similar to that of the videocomposers described in a preceding section, and thus will permit online page composition prior to printing. The most complex systems will merge graphic and textual input. Single or multiple copies of predefined pages will be printed, xerographically or by other means, by computer-controlled equipment. Microfilm output also will be possible. If intended recipients of a given document are located at geographically scattered sites, the text will be transmitted electronically to computer-driven printers at those sites. While many of these operations are possible now using carefully selected components and a variety of interface devices, a fully developed CAMIS configuration will perform successive information-processing tasks automatically. Any required interfaces will be transparent to the terminal operator.

The advantages of CAMIS technology are especially significant in those offices with high-volume printing requirements. Under computer control, the style, quantity, time, and place of printing can be specified by the operator to meet the requirements of individual applications. Much of the physical handling, inventorying, and related tasks associated with large-scale document reproduction operations either will be reduced or eliminated significantly. Systems will be designed for ease of use by clerically trained employees, thus eliminating the trade-craft skills required during prior generations of printing technology. Short-run printing will become economically feasible, and customized publications—tailored to narrowly defined audiences—will be assembled from broad accumulations of machine-readable text. From the standpoint of the document creator, CAMIS technology will integrate the pre-press and printing operations that have historically separated authors from those worksteps involved in the reproduction and final presentation of the documents they generate. The office-based document originator therefore will be able to exercise much greater control over the final form and dissemination of paper information.

Given the recent rapid pace of change in reprographic technology, it is likewise that the implementation of partial or full CAMIS configurations will be feasible by the mid-1980s. Many CAMIS concepts depend, however, on developments in the field of electronic communications, which is discussed in the next chapter.

6
Electronic Mail and Message Systems

Electronic mail systems • Image-oriented message transmission • Character-oriented message transmission

As its title indicates, this chapter deals with those information-processing systems that are designed for the transmission of written messages from one point to another electronically rather than by physical delivery.[1] As such, electronic mail and message systems (EMMS) offer a specialized capability within the broad field of electronic information communication, which includes voice as well as nonvoice messages.

The present scope of electronic mail and message systems is limited to written communications which have been generated by hand, typewriters, typesetters, or any of the several terminal devices described later in this chapter. It is possible, however, that the projected development of "speech mail" systems, in which computers will be used to store and forward digitized voice communications, will necessitate a broadening of the scope of EMMS to include noninteractive spoken messages. The intent of the present limitation is to exclude the most common class of interactive voice communications—telephone calls. It is likewise important, for purposes of this discussion, to distinguish the electronic transmission of messages—that is, communications between individuals or organizations—from the electronic transmission of data. The latter field is technically concerned with the transmission of information in the form of encoded characters generated by computers, terminals, or related devices. As such, some of the electronic mail and message systems discussed in this chapter are designed to transmit computer-generated, character-coded data, but most data communication applications involve the transmission of information other than messages. This is the case, for example, with regard to most terminal to computer communications encountered in the online information systems described in Chapter Three. On the other hand, some electronic mail and message systems are designed to transmit document images rather than character-

153

coded data. Thus, the fields of electronic message communication and electronic data communications are related, but not identical.

The socio-economic significance of reliable message delivery capability is reflected in the fact that all developed countries have nationalized mail service.[2] While such services often are criticized as ineffective, they generally have taken advantage of technological innovations in ground and air transportation to achieve improvements in delivery speed. Yet, there is necessarily a lower limit on the speed of physical mail delivery. In the continental United States, for example, the theoretical lower limit for coast-to-coast delivery of a first-class letter is one-half day, but actual delivery rarely requires less than three days and may take much longer. While many office applications can be satisfied with such speeds, the need for faster delivery of time-sensitive messages is addressed by the various electronic transmission technologies described in this chapter: (1) facsimile and other image-oriented technologies; (2) the telegraph-type services, TWX and telex; (3) Mailgram and related services which combine electronic transmission with physical delivery; (4) communicating word processing; and (5) communicating optical character recognition equipment. Computer-based message systems, which employ electronic transmission but have a somewhat different rationale, likewise are discussed.

Some electronic mail and message systems—notably facsimile transmission and the telegraph-type services—are based on technologies that have been available for several decades but recently have become the focus of intense office interest. This interest is based on a combination of managerial considerations, application characteristics, and technological developments. From the managerial standpoint, there is growing awareness of the inherently wasteful nature of much time spent in travel, accompanied by a recognition that electronic communications often can be substituted successfully for travel.[3] This managerial consideration relates to a point previously noted in Chapter One—that is, the potential for productivity improvements inherent in the restructuring of prevailing work patterns through the application of information-processing technology. From the application standpoint, the widespread acceptance of word processing has resulted in the existence of machine-readable representations of many office messages. Such machine-readable messages lend themselves readily to electronic transmission. Other important application considerations include the increasingly international character of much office work, which has stimulated a need for the delivery of messages to areas of the world which cannot be addressed by conventional physical delivery in a timely and reliable manner, and the need for high-speed transmission of messages which pertain to data, much of which are transmitted electronically. In applications involving computer-based message systems, discussed later in this chapter, the time-sensitive nature of the transmitted messages is often a secondary consideration. Such systems typically are implemented to extend the span of supervisory control, to eliminate time-consuming telephone calls, or to simplify the distribution and subsequent handling of messages.

Interest in electronic mail and message systems likewise has been stimulated by recent developments in communications technology. With the exception of the network of telegraph lines used for TWX and telex communications, electronic transmission in the United States historically has utilized the telephone network—the existing system of switched and leased lines which links a geographically scattered area and thus constitutes the prevailing national communications infrastructure. While it provides a variety of features and excellent geographic coverage, several factors limit the utility of the public telephone network for electronic mail and message transmission:

1. The telephone network originally was designed for the transmission of the analog signals characteristic of voice communications rather than the digitally coded signals commonly employed in the electronic transmission of messages and data. Such signals, as previously noted in the discussion of terminals presented in Chapter Three, must be converted to analog form, via modems, prior to transmission.

2. The telephone network consists largely of copper wires which are vulnerable to transmission impairments and must operate at relatively low speeds. Most telephone lines are described consequently as "narrow band" communication facilities, reflecting the limited range of signalling speeds they will support. Public-switched telephone lines are, for example, theoretically incapable of accurate transmission at rates greater than 4800 bits per second. In actuality, however, transmission over such lines typically occurs at speeds of 1200 bits per second or lower. Transmission rates up to 9600 bits per second are possible with specially conditioned leased lines, while several leased lines can be grouped to create "wide band" communication facilities capable of signalling rates in excess of 50,000 bits per second. Such high-speed facilities are, however, expensive and only can be justified in applications involving substantial volumes of transmission between two points.

3. Charges for public-switched telephone service are incurred on the basis of both elapsed time and the distance between communicating parties. This rate structure penalizes long-distance message transmission.

The existing communications infrastructure is not, however, a static entity. Within the framework of the telephone network, for example, the Bell System offers leased digital lines which can support transmission at speeds up to 1.544 million bits per second. Modems are not required and, being digital, the transmitted signal is less vulnerable to the impairments typically encountered in analog communications. The Dataphone Digital Service (DSS) is currently available in most large and many medium-size cities. As another indication of the improvements being made in the existing telephone network, the American Telephone and Telegraph Company currently is converting telephone lines from copper wires to fiber optics in the northeastern "corridor," which extends from Washington, D.C. to New York City to Cambridge, Massachusetts. Fiber optics, which have replaced conventional telephone wires in other parts of the United States and Canada, conduct digitally coded signals in the form of pulses of light. Compared to copper wires, fiber optics offers the advantages of relatively low signal losses over long distances, very high transmission speeds, compact size, immunity to external electromagnetic interference, and a plentiful supply of raw materials needed to create the required glass fibers. From the users' standpoint, these advantages should be reflected in improved communications capabilities better suited to electronic message transmission.[4]

Figure 6-1. Fiber optics conduct digitally coded signals in the form of pulses of light. Compared to copper wires, fiber optics offer the advantage of low signal loss, high transmission speed, compact size, and immunity to electromagnetic interference. The optical waveguides shown here are specifically designed for such short-distance applications as the interconnection of computers and peripheral devices. (Courtesy: Corning Glass Works)

In terms of electronic mail and message systems, however, the most significant recent developments in communications technology have involved the telephone network indirectly, if at all. The discussion of online information systems in Chapter Three made reference to the important role played by the so-called value-added carriers—Telenet and Tymnet, for example—in data communications. These carriers are playing an increasingly important role in message communications as well. Briefly, the value-added carriers provide nationwide networks of analog lines, leased from the telephone company, which link computers, terminals, and various other data communications devices. While geographic coverage and utilization procedures vary from vendor to vendor, users in large- and medium-size cities typically access these value-added carriers through their local telephone networks. But, unlike the conventional telephone network which establishes a fixed circuit for the duration of a transaction, the value-added networks use "packet switching" technology to subdivide a given communication into segments, each of which may be transmitted to its intended destination by a different route. Operating under computer control, the network reassembles the segments or "packets" into the proper sequence at the destination point. Unlike the telephone network, transmission charges are based on elapsed time without regard to distance and typically compare very favorably with conventional long-distance telephone rates. In addition to permitting lower cost intra-company communications, the value-added networks provide time-sharing service bureaus and online data base services with the economical communications facilities essential to the establishment of a nationwide customer base.[5]

While most often used for data communications in the broadest sense, most value-added networks support some form of electronic mail and message communications. As discussed later in this chapter, Tymnet and Telenet offer computer-based message systems, and both networks provide access to time-sharing service bureaus and other companies that offer electronic message services. Telenet, for example, is the long-distance communication facility for the COMET electronic message system developed by Computer Corporation of America and for the HERMES system marketed by Bolt, Berenak, and Newman. Similarly, several value-added networks specialize in electronic message transmission. Examples include ITT Faxpak and SPC Speedfax, two carriers that offer facsimile transmission services, and Graphnet, a value-added network which permits communication among TWX, telex, and facsimile terminals.

The value-added carriers employ innovative transmission techniques which provide greater error resistance and a rate structure that favors long-distance communications. The transmission facility, however, consists of leased telephone lines which are subject to the speed limitations described earlier in this chapter. Much transmission consequently occurs at speeds of 2400 bits per second or lower. As discussed in previous chapters, many of the output devices which are available for use in electronic mail and message systems—including display terminals, newly developed laser-oriented xerographic printers, and even some conventional impact printers—are capable of operating at speeds greater than the wire-based value-added networks can reliably support. The resulting transmission "bottleneck" has stimulated interest in three alternative communication technologies: microwave, satellites, and coaxial cable.

While most often applied in fields such as radar and radio astronomy, terrestrial microwave transmission facilities have been used for voice and data communication for several decades. Such facilities are available in most major metropolitan areas and are utilized in many office-related applications. Some organizations have established private microwave links. MCI Communications is perhaps the best known of the public terrestrial microwave carriers.

Utilizing radio as an alternative to wires, microwave signals may travel at speeds up to several billion bits per second. Actual transmission rates vary with the particular radio frequency utilized, but even at the lowest rates, the speed of microwave transmission is at least 20 times greater than the fastest available conventional telephone facilities. Terrestrial microwave transmission facilities are not, however, as versatile or as wide ranging as the telephone network. To prevent deflection by obstacles in the communication path, microwave signals are transmitted by a series of directional antennas spaced at distances of 20 to 30 miles. In large cities, for example, these antennas typically are located on roofs of tall buildings. The requirement for an unobstructed communication path necessarily limits the application of terrestrial microwave to relative short-distance transmission where the establishment of a "line of sight" between the antennas is possible. As a further constraint on the application and continued development of terrestrial microwave communications, there is considerable congestion of currently allocated microwave frequencies in major metropolitan areas.

These problems of distance and congestion can be minimized or overcome, however, if microwave signals are transmitted to geosynchronous satellites for amplification and retransmission to earth stations located at or near the intended reception point. Satellite communication thus constitutes a special class of microwave transmission.[6] A geosynchronous satellite is so called because throughout its orbit it maintains its position relative to a given point on the earth's surface. It thus serves as an apparently stationary relay device in the sky. Such satellites, more than 70 of which are now in orbit, have been used for television and international telephone transmission since the 1960s. While developing countries have long recognized the potential significance of satellites for domestic voice and data communications, American interest in such domestic satellite communication is comparatively recent. However, it has been stimulated by increasing awareness of the inadequacies of existing terrestrial communication facilities, as previously discussed. Since the first WESTAR satellites were launched by Western Union in 1974, avaliable domestic satellite facilities have been well publicized. The WESTAR satellites, for example, provide data, voice, and video transmission to 20 cities. Similarly, American Satellite Company, which owns a portion of the WESTAR satellite system and is itself owned by Continental Telephone and Fairchild Industries, is a well-established domestic communications carrier which leases satellite channels, earth stations, and related support equipment. The typical customer begins with 2 channels capable of transmitting at 56,000 bits per second each and expands capacity gradually as requirements dictate. The leased satellite channels commonly are used for a combination of voice, data, and facsimile transmission. American Satellite also will support video conferencing and will connect to international satellite or terrestrial carriers.

While American Satellite has made its communication facilities available for some time, two recently announced satellite services—Satellite Business Systems (SBS) and the Xerox XTEN satellite network—have been discussed more widely in the context of office automation. Of the two, the Xerox XTEN network was still in development at the time this chapter was written. It will offer all-digital data and facsimile transmission at speeds up to 256,000 bits per second using leased satellite circuits. Microwave transmission will be used to link shared earth stations with rooftop antennas located on the customer's premises. Satellite Business Systems—a joint venture of IBM, Aetna, and Comsat General—launched its first satellite in the fall of 1980. Documents filed with the Federal Communications Commission at that time outline several levels of service to be initiated in 1981 and 1982. A private network service is designed for large corporations and government agencies with three or more on-premises earth stations and a requirement for high-volume computer-to-computer data transmission, high-speed document transmission via facsimile, and video teleconferencing. Digital transmission, at speeds up to 224,000 bits per second, is used for all communications.

To support high-volume electronic mail delivery, Satellite Business Systems has contracted with AM International for the development of a digital facsimile transceiver capable of transmitting a letter-size document in as little as five seconds. As discussed later in this chapter, other available facsimile equipment cannot transmit a letter-size document in less than 20 seconds. The device developed by AM International is further notable for its use of xerographic technology to produce high-quality paper copies. It is expected that the typical private network user will have 25 earth stations. For lower volume users, a shared service—initially designed for digitized voice transmission only—combines on-premises and shared earth stations. Data transmission capability, at speeds up to 56,000 bits per second, will be added later, as will interconnections with the switched telephone network, value-added networks, and other terrestrial carriers.

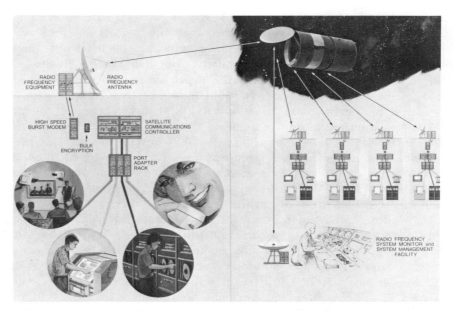

Figure 6-2. Satellite transmission represents an important alternative to conventional wire-based communication facilities. The SBS network is designed for the high-speed transmission of digitally-coded voice, data, and image communications. Data can be transmitted at speeds ranging up to 6.3 million bits per second. Earth stations, located on the customer's premises, reduce or eliminate dependence on terrestrial transmission links. (Courtesy: Satellite Business Systems.)

The terrestrial microwave and satellite technologies described above provide wideband transmission facilities designed for high-speed, long-distance communication. Coaxial cable is increasingly viewed as a high-speed alternative to conventional telephone lines for local communications at speeds up to 56,000 bits per second.[7] Long associated with the provision of television transmission to remote areas, coaxial cable has been used for environmental, fire, and security control in various manufacturing and military applications. More significant for purposes of this discussion, however, is its use as the transmission medium for data or video communications in single or multi-building local networks. Such facilities have been installed, for example, by many corporations, government agencies, universities, and research facilities. Several vendors are in the process of developing coaxial-based office networks which will support the connection of various types of information-processing equipment, including word processors, terminals, computers, and "intelligent copiers." The most widely

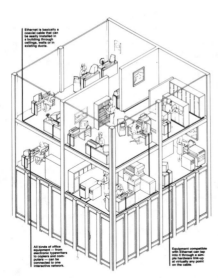 

Figure 6-3. Conventional coaxial cable is increasingly viewed as a high-speed alternative to conventional telephone lines for local communications. Ethernet, the most widely publicized coaxial cable network, is designed to permit the interconnection of computers, word processors, terminals, intelligent copiers, and related equipment. The cable itself is installed in walls, ceilings, or ducts (left) and individual information processing machines are connected to it via specially designed outlets (right). (Courtesy: Xerox Corporation)

publicized of these networks, Ethernet®, is being developed by Xerox in conjunction with the Digital Equipment Corporation, a well-known manufacturer of minicomputers and related peripheral equipment, and Intel, a manufacturer of microprocessors and related electronic circuitry. Xerox has announced the availability of two devices for attachment to the Ethernet® network and has revealed its intention to license other manufacturers to develop compatible products. At the time this chapter was written, Ethernet® specifications had been submitted to the Institution of Electrical and Electronic Engineers for consideration as an IEEE standard for local networks. Other information system manufacturers, however, are working on alternative networks. For example, Zilog, a microprocessor manufacturer that is a subsidiary of Exxon Corporation, is developing its own coaxial-based network called ZNet. It will interconnect various microprocessor-based products, including terminals, word processors, and microcomputers.

Taken together, the technological developments described in the preceding paragraphs will increasingly provide a foundation for the widespread use of electronic transmission as an alternative to the conventional physical delivery of messages. For purposes of the following discussion, electronic mail and message systems can be divided into the two broad groups mentioned at the beginning of this chapter: those that transmit messages in the form of document images; and those that transmit the information content of messages as sequences of encoded characters. The next section begins with a discussion of the most commonly encountered form of image transmission—facsimile.

IMAGE-ORIENTED MESSAGE TRANSMISSION

Facsimile Technology

Facsimile—often simply abbreviated as "fax"—is (1) an exact copy of a document or (2) the process or result of the process by which fixed graphic images are scanned, transmitted electronically and reproduced either locally or remotely.[8] As is true of other electronic mail and message technologies discussed later in this chapter, a facsimile system consists of an integrated configuration of equipment and communication facilities designed to achieve the transmission of information-bearing signals from one point to another. The transmission itself originates at a source terminal called a *transmitter;* it terminates at a sink terminal or *receiver.* Typically, transmission and reception capabilities are united in a single device called a *transceiver.* The facsimile signal is conveyed by a communication medium which, in most applications, is a telephone line.

The function of the facsimile transmitter is to scan a document (called the *subject copy*), using an optical assembly and light source to convert the document's tonal variations into a pattern of varying electrical signals suitable for transmission over the intended communication medium. In the oldest, and still widely used, method, the subject copy is wrapped around a cylinder which rotates at high speed while a scanning mechanism moves down the long axis of the subject copy. In flat-bed scanning, a technique increasingly used in newer facsimile equipment, the subject copy is inserted into a slot from which it is transported past a fixed scanning mechanism. Acceptable input sizes vary, a maximum of 8.5 by 14 inches being most common in rotating cylinder equipment. Flat-bed scanners may accept longer documents and are less affected by variations in document thickness.

Regardless of the input method, the facsimile transmitter contains a light source and an optical assembly consisting of a photocell capable of analyzing light reflected from the subject copy. The transmitter divides the subject copy into areas called *picture elements* or *pixels,* the number and size of which varies with the scanning resolution, as discussed later in this chapter. Most scanners employ mechanical illumination methods with conventional incandescent spot or flood lights. Fiber optics and laser scanners, however, are used in several newer transmitters, and the popularity of such electronic scanning mechanisms is expected to increase.

163

Figure 6-4. The typical facsimile transceiver is a desk-top device which uses an optical assembly and light source to convert a document's tonal values to an electrical signal suitable for transmission over telephone lines or other communication facilities. Special features vary from model to model. The Qwip 2150, for example, can transmit a letter-size document in 2, 3, 4 or 6 minutes; includes an automatic document feeder; and features unattended telephone answering. (Courtesy: Exxon Office Systems Company)

Whether mechanical or electronic illumination is employed, however, light reflected from a given picture element is measured by a photocell which then generates an electronic signal, the intensity of which varies with the intensity of the reflected light. Most office documents consist of dark text on a white or other light background. Dark areas of the document will absorb light, while the white areas will reflect it. The facsimile transmitter generates an electrical signal which may reflect these light intensity variations in either analog or digital form. Most facsimile transmitters are analog devices whose output consists of a continuously varying electrical signal which represents the subject copy as a serial stream of light intensity variations. Depending on the transmitter's sensitivity, varying intensity levels may represent eight or more shades of grey. Digital transmitters, on the other hand, represent dark and light picture elements by a series of discrete on/off electrical pulses rather than a continuously varying signal. In examining successive pixels, the digital transmitter makes a determination as to whether each should be represented as black or white in accordance with a pre-established threshold of light reflectance value. Grey areas will be transmitted as either black or white, depending on their intensity. Digital transmitters, however, do permit the application of encoding techniques which can reduce transmission time significantly. Those techniques are discussed later in this chapter.

As previously noted, the communication medium in most facsimile applications is a telephone line. In fact, the ability to interconnect with the public telephone network has contributed significantly to the growth of office-based facsimile applications. As described above, the facsimile transmitter generates an analog or digital signal which represents the light reflectance values of picture elements successively encountered in the subject copy. Whether analog or digital representation is used, the resulting electrical signal—sometimes called the *baseband signal*—cannot be transmitted directly over telephone lines intended for voice communications. The public telephone network, as discussed earlier in this chapter, is designed to respond to the range of frequencies represented in the human voice. That range is nominally 300 to 3,300 cycles per second (or *hertz*), and the bandwidth of a typical telephone line is, consequently, 3,000 hertz. Frequencies represented in the baseband signal may range from a high of several thousand hertz to a low of zero or near zero. White areas of the subject copy may generate, for example, a very low frequency signal which cannot be transmitted over long distances using telephone wires. The resulting transmission problem, however, can be overcome through modulation—a process whereby the baseband facsimile signal is shifted upward into the range of frequencies within which voice grade telephone lines respond.

The concept of modulation was introduced in Chapter Three where it was applied to the conversion of the digital signal generated by a computer terminal to the analog form required for transmission over telephone lines. In digital facsimile systems, this form of modulation is likewise required, but modulation must be applied to baseband analog signals as well. Through modulation, an information-bearing baseband signal is impressed on a specially selected carrier signal that lies within the range of frequencies to which telephone circuits respond. The information represented by variations in the baseband signal is communicated by altering one of three characteristics of the carrier signal. In frequency modulation (FM), which is used by the majority of installed facsimile equipment, the information content of the baseband signal is represented by changes in the frequency of a carrier of fixed amplitude. In amplitude modulation (AM), which is used by certain facsimile devices manufactured by Graphic Sciences, the information content of the baseband signal is represented by changes in the relative loudness of a carrier signal of fixed frequency. In phase modulation (PM), a method used by some high-speed facsimile transceivers, the information content of the baseband signal is represented by changes in the time when cycles of a fixed frequency occur. As discussed later in this chapter, the selection of a particular modulation technique has a significant impact on the compatibility of facsimile equipment. Regardless of the technique employed, modulation is varied through a modem which is typically built into the facsimile transceiver. Some facsimile transceivers feature an acoustic coupler as an alternative to a conventional modem. Acoustic coupling is, of course, essential in portable facsimile equipment.

As the sink device, the function of the facsimile receiver is to decode (demodulate) the transmitted signal, interpreting its intensity variations and recording them as a pattern of light and dark areas that form a paper facsimile of the subject copy. The receiver thus can be visualized as a remote copier. Facsimile receivers resemble transmitters in appearance and construction. As previously noted, most manufacturers combine transmission and reception capabilities in a single transceiver device. If full duplex capability is provided, the transceiver will be able to transmit and receive simultaneously. If half duplex capability is provided, the two functions cannot be performed simultaneously. While transceivers are generally less expensive and more convenient, a combination of discrete transmitters and receivers may offer greater flexibility in special applications. Regardless of the particular equipment configuration, several recording processes are in current use:

1. The *electrolytic* process—the oldest facsimile recording technology—features a thin paper saturated with an electrically conductive solution. A current of varying voltage then is passed from a stylus through the paper, discoloring those areas that correspond to dark picture elements in the subject copy.

2. The *electropercussive* or impact process, likewise an older recording technology, creates the facsimile copy by activating a stylus that strikes a carbon-impregnated paper in areas corresponding to dark picture elements in the subject copy.

3. In the *electrosensitive* or *electrothermal* process, a special recording paper, which has a dark carbon underlayer covered with an opaque white coating, is used. An electric current, passed from a stylus through the paper, decomposes the coating in areas corresponding to dark portions of the subject copy, revealing the dark underlayer.

4. In the *electrofax* or direct electrostatic process, as described in Chapter Five, a photoconductive paper coated with zinc oxide is charged in darkness and exposed to a light source capable of displaying a pattern of light intensity variations corresponding to the various picture elements in the subject copy. The resulting latent image of charges is made visible through the application of toner.

5. Similar to the electrofax process, the familiar *xerographic* process is a variant form of electrostatic recording. Charges, exposed on an intermediate surface by a laser or other light source, are developed by toner, and the resulting image is transferred to an ordinary sheet of paper. Being uncoated, the resulting facsimile copy typically enjoys high user acceptance.

In addition to the processes described above, other recording technologies may be utilized in facsimile receivers designed for special applications. Facsimile equipment developed for the transmission of photographs and fingerprints between criminal justice agencies, for example, typically utilizes silver halide recording to achieve high quality images. For the transmission of news photographs from wire services to newspapers, laser-based recorders generate high resolution images on thermal silver paper. Such special-purpose facsimile systems, however, are not utilized commonly in office applications.

Performance Characteristics

Facsimile is a comparatively old technology. For more than 50 years, it has been used by newspapers, police departments, weather stations, and the military for the transmission of time-sensitive documents. It has long been considered a technology with obvious business utility and, given the currently intense interest in high-speed communications discussed earlier in this chapter, it is enjoying considerable popularity in office applications. Through the early 1970s, however, the acceptance of facsimile technology in general office applications was impeded by problems of equipment reliability, output quality, transmission speed, and system compatibility. These problems have been resolved, for the most part, in the use of newer equipment.

To set the problem of facsimile reliability in perspective, it is important to note that no communication system is immune to transmission errors. Messages transmitted via the U.S. Postal Service or other physical delivery methodologies are, for example, subject to missorting, mutilation, or destruction. Telephone lines are likewise subject to a number of transmission impairments, including crosstalk, echo, and external interference. Facsimile and other electronic message delivery systems introduce the additional potential for equipment malfunction, but it is important to note that facsimile is typically less sensitive to such malfunction than the other electronic message technologies discussed later in this chapter. Rather than resulting in the obliteration or improper transmission of entire characters, facsimile errors typically manifest themselves in appearance blemishes in portions of the recorded output. The facsimile copy itself remains usable. Nevertheless, facsimile users during the 1960s and the early 1970s often reported incidents of output quality degradation due to equipment or transmission line malfunctions. In some cases, transmission was initiated without adequate synchronization of transmitter and receiver. In others, telephone line problems resulted in unusable output, necessitating retransmission. To minimize the potential for such errors, facsimile transceivers now routinely incorporate synchronization logic to insure that transmitting and receiving units are in phase. Similarly, most facsimile receivers allow the operator to view output as it is being recorded. If unusable copy is being received, the operator can activate a special signal instructing the transmitter to terminate transmission. Some facsimile equipment continuously will monitor line connections throughout transmission. In the event of telephone line problems, transmission is terminated and an alarm is activated.

It is possible that much of the reported dissatisfaction with facsimile in office applications stems from a misunderstanding of the role of output quality in facsimile systems. Unlike other reprographic technologies—notably, micrographics and full-size copying/duplicating—in which the highest attainable output quality is invariably desired, facsimile recording technologies are oriented typically toward the creation of useful—that is, legible—output. Two levels of legibility typically are distinguished: character legibility, in which each individual character is readily identifiable; and word legibility, in which the identities of words can be established from surrounding context, although some characters may not be individually identifiable. While character legibility is required for the accurate interpretation of numerical data, labelled graphs, and tables, word legibility suffices for most textual documents.

The attainment of an appropriate level of facsimile output quality depends on several interrelated factors, including the scanning resolution and the nature of the subject copy. Unlike micrographics and other reprographic technologies in which resolution is a measurement of accomplished output quality, facsimile resolution is an expression of the potential for quality output in terms of the number of scan lines in a given linear dimension of the subject copy. Facsimile resolution is typically expressed in *lines per inch* (1pi) and is measured both horizontally and vertically. As discussed earlier in this section, the subject copy and the facsimile copy can each be visualized as a grid of picture elements, the exact number of which will depend on the number of scan lines. In a system offering a resolution of 100 scan lines per inch both horizontally and vertically, a letter-size (8.5- by 11-inch) subject copy would be divided into ((11 × 100) × (8.5 × 100)) or 935,000 picture elements, each of which would be analyzed for its light reflectance value. The greater the number of picture elements, the sharper the transitions from light to dark areas will appear in the facsimile copy. Since the mid-1960s, the most widely available business facsimile systems have provided maximum horizontal and vertical resolution capabilities in the range 88 to 100 lines per inch, with 96 lines per inch being the most common. Such resolutions are quite satisfactory for the transmission of most typewritten office documents. Resolutions in the above range are adequate for word legibility in the transmission of printed text set in type as small as eight points, but are of marginal or no utility for the transmission of numerical or tabular data, labelled graphs, or footnotes set in six point or smaller type faces. Where such materials must be transmitted, several available business facsimile transceivers offer resolutions up to 200 horizontal and vertical lines per inch. The resulting improvement in output quality, however, typically is attained at the expense of transmission speed, as discussed below.

The operating speed of facsimile equipment is expressed in terms of the time required to transmit a letter-size (8.5- by 11-inch) document. Through the mid-1970s, the typical facsimile transceiver was an analog FM device capable of transmitting a letter-size document in 6 minutes at 96 lines per inch vertical and horizontal resolution. Alternatively, the operator could reduce the transmission time for a letter-size document to 4 minutes with a corresponding degradation in resolution to 96 lines per inch (horizontal) by 64 lines per inch (vertical). Many office documents are able to tolerate such degraded resolution without a loss of utility in the facsimile output. The reduction in transmission time results, of course, in lower telephone charges in applications utilizing the long-distance dial-up network. While the bulk of installed facsimile transceivers are still of the 6/4-minute FM analog type, most newer analog transceivers are capable of transmitting a letter-size document in 3 minutes at 96 lines per inch vertical and horizontal resolution, or in 2 minutes at 96 by 64 lines per inch. The most significant recent increases in transmission speed have resulted, however, from the development of digital facsimile equipment. As discussed earlier in this chapter, digital facsimile transmitters convert the light reflectance values of successive picture elements in the subject copy to binary pulses representing either light or dark areas. The resulting digital signal then is analyzed and encoded by a microprocessor prior to transmission to reduce (compress) the number of pulses necessary for the receiver to reconstruct the subject copy. The compressed signal, however, must be converted to analog form, via a modem, for transmission over the voice-grade telephone network.

The prevailing data compression methodology in digital facsimile systems is called *run-length encoding* because it results in a coded message representing the run-lengths of successive black and white picture elements of the subject copy.[9] Consider, for example, a digital signal resulting from the analysis of a facsimile scan line in which the first eight picture elements are:

W W W B B B B B

where W signifies a white area and B a black area. Using run-lengths, this signal can be encoded as:

3 W 5 B

thereby reducing the amount of information to be transmitted. Because the polarity of picture elements alternates between two extremes, however, it is only necessary to know whether the first element is white or black. The signal thus can be compressed further to:

3 W 5.

Figure 6-5. Digital facsimile transceivers utilize data compression methodologies to drastically reduce the time required for document transmission. Available models incorporate automatic feeders and other features designed to minimize operator involvement. (Courtesy: 3M Company.)

In the above example, run-length encoding results in a compression factor of 8/3 or 2.7. Even when additional control bits are added to the encoded facsimile signal, required transmission times will be markedly reduced when compared with an uncompressed analog facsimile signal. Available digital facsimile transceivers can transmit a letter-size document in 20 to 90 seconds, depending on the desired resolution and the communication facility utilized. In most cases, data compression techniques are accompanied by *variable velocity scanning* in which the flat-bed transmitter looks ahead to identify blank lines and other white areas of the subject copy that can be scanned more rapidly than denser text areas.

Because data compression is accomplished by an integral microcomputer, digital facsimile transceivers are typically three to five times more expensive to rent or purchase than their slower analog counterparts. In every application, however, there is a break-even point (in terms of volume of transmitted documents) where the accumulated savings in long-distance telephone charges will outweigh the higher equipment cost, resulting in a lower unit cost per document transmitted. Beyond that break-even point, which will necessarily vary from application to application, the cost advantage of digital versus analog facsimile increases with transmission volume.

Whether digital or analog, all business facsimile transceivers incorporate simplified controls and features designed to minimize operator decision making. As a result, most facsimile equipment can be operated by ordinary office personnel rather than by specially trained equipment operators or similarly classified employees. Facsimile equipment manufacturers are increasingly emphasizing standard or optional features designed to minimize personnel involvement in the transmission process. *Automatic feeding,* as the name implies, provides for the automatic, high-speed mounting or insertion of successive subject copies in cylinder or flat-bed transmitters, respectively. Automatic feeding can be useful in reducing the labor involvement and long-distance telephone charges associated with the time required to change subject copies in multi-page transmission transactions. Subject copies to be loaded are inserted in a hopper, the capacity of which ranges from 50 to 100 documents, depending on the model. Most automatic feeders will accept documents of mixed sizes and thicknesses.

At the receiving terminal, the counterpart of automatic feeding is *automatic answering* and *unattended reception.* An automatic answering device eliminates operator involvement in the establishment and maintenance of telephone contact at the facsimile receiver. Unattended reception usually is facilitated by roll-feed paper mechanisms with capacities up to 500 copies. Depending on the model, received facsimile copies may be cut to size and stacked in a tray or left in continuous roll form. Following reception of the last document, the automatic answering device terminates the telephone connection and returns the transceiver to the standby condition. Automatic feeding and unattended reception are typically most significant in applications utilizing high-speed facsimile equipment where they minimize time wasted in the establishment of the telephone connection and the changing of subject copies in multi-page transactions.

To enable offices that close at or before five o'clock P.M. to take advantage of evening long-distance rate reductions, several facsimile equipment manufacturers offer an optional *automatic dialer*—a microprocessor-controlled device with a digital clock and a touchtone keyboard which can store telephone numbers to be dialed at designated times. A single document may be premounted or preinserted for automatic transmission. More commonly, a stack of documents is preloaded in an automatic feeder, and telephone numbers and transmission times are entered via coded cards or the operator keyboard. Capacity varies from model to model, with some dialers capable of storing as many as several hundred telephone numbers. Automatic dialing can be especially useful in facsimile transmissions between offices located in different time zones. Where offices at both the transmitting and receiving points will be closed, automatic dialing can be used in conjunction with unattended reception. In a variant form of automatic dialing called *polling,* an appropriately equipped facsimile transceiver can call other transceivers and, using a special signal, query (poll) them to determine whether they have been preloaded with documents to be transmitted to the polling transceiver. Thus, polling allows the receiving office to control the time of transmission.

Compatibility

Compatibility refers to the ability of a given facsimile transmitter and receiver, or two transceivers, to communicate with one another. In business applications involving intracompany communications, centralized procurement policies and practices can insure the compatibility of facsimile equipment in geographically scattered offices. But widespread acceptance of electronic mail and message transmission requires the ability, however occasional, to communicate outside of a closed system.

Facsimile equipment compatibility depends on several technological factors, including operating speed and modulation technique. To facilitate the transmission of document images among the equipment of different manufacturers, international facsimile standards have been developed by the Comité Consultatif International Télégraphique et Téléphonique (CCITT).[10] As previously discussed, the facsimile transceivers developed through the mid-1970s were designed to transmit a letter-size document in either four or six minutes, depending on the selected resolution. CCITT standards for four/six-minute transceivers, however, were developed too late to affect the installed base of such equipment in the United States. Consequently, offices employing the equipment of one manufacturer cannot invariably transmit to offices employing the equipment of another manufacturer, even though the devices are nominally identical in operating speed and modulation technique. The FM analog transceivers of most manufacturers may, for example, be compatible in the six-minute mode, but they are rarely compatible in four-minute transmissions. As a further complication, those devices which employ amplitude modulation, as previously discussed, cannot communicate with FM analog transceivers, regardless of speed. Compatibility is fortunately improved with newer facsimile equipment. CCITT standards have been developed for three-minute analog transceivers and for high-speed digital transmission. Widespread adherence to these standards greatly simplifies the task of facsimile system design.

Still, the continued utilization of many different types and generations of facsimile transceivers makes the ability to communicate with the widest range of installed equipment essential to the acceptance of facsimile as a viable form of message delivery. Recognizing this, equipment manufacturers have emphasized multi-speed capabilities in recently introduced models. Thus, some essentially AM transceivers can be optionally configured for FM transmission, while some high-speed digital units also can communicate with the installed base of slower FM analog units. It is a measure of the complexity of the facsimile marketplace, however, that even the most versatile multi-speed transceivers remain incompatible with some available devices.

One of the most promising developments in facsimile equipment compatibility involves the value-added networks discussed earlier in this chapter. As previously explained, the value-added facsimile networks—such as Faxpak, developed by International Telephone and Telegraph, and Speedfax, a service of Southern Pacific Communications—offer low-cost transmission facilities as an alternative to the conventional long-distance telephone network. These value-added networks offer software which will enable otherwise incompatible facsimile transceivers to communicate with one another. Such software, which is now available on a limited basis, provides for the transmission of facsimile messages to a central computer that translates them for retransmission to designated receivers. One of the value-added networks—Graphnet, a service developed by Graphic Scanning Corporation—enables TWX and telex users to communicate with facsimile receivers. A similar service recently was announced by Tymnet as part of its On-Tyme electronic mail system discussed later in this chapter.

Microfacsimile

In providing communication capabilities between keyboard-based message systems and facsimile equipment, value-added carriers such as Graphnet are extending the range of acceptable facsimile input to include machine-readable data as well as paper documents. Microfacsimile, as the name implies, denotes the technology and equipment designed for the transmission of images from microforms to remote locations where they may be displayed on a screen or recorded on paper or microfilm. Microfacsimile thus links electronic mail and micrographic storage systems, just as COM and CAR link micrographics and computer technology. While the concept of microfacsimile has been discussed for several decades, an office market for such systems generally has failed to materialize. Historically, the ease and economy with which microforms can be duplicated and stored in multiple locations has obviated the need for remote transmission of microimages. Through the late 1970s, only the Alden Electronic and Impulse Recording Company has offered commercially available microfacsimile equipment. In addition, several one-of-a-kind systems have been developed for military applications.

There are indications, however, of recently renewed office interest in microfacsimile, and several interesting systems have been developed over the past few years. The Telefiche System, developed by Planning Research Corporation, is described here as representative of the current state of the art in microfacsimile. The system consists of an integrated configuration of micrographic, computer, and data communication components designed to permit the transmission of a document image from a central file of roll microfilm or microfiche and its reconstruction as either a screen display or

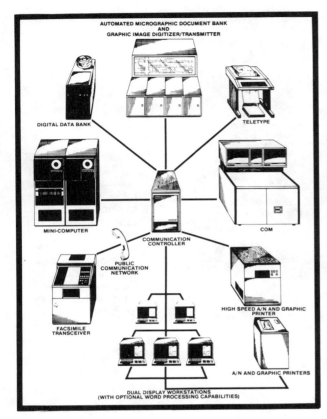

Figure 6-6. Microfacsimile, as the name implies, permits the transmission of a document image from a central file of roll microfilm or microfiche and its reconstruction as either a screen display or a paper copy at a remote location. The Telefiche System, developed by PRC Image Data Systems Company, links a micrographics data base with a wide range of peripheral devices and remote work stations. (Courtesy: Planning Research Corporation.)

a paper copy at a remote location. As is true of conventional facsimile systems, the Telefiche System includes a scanner which—depending on the particular configuration selected—will accept either 24X, 16-mm roll microfilm, or 24X or 48X microfiche. Other roll and fiche formats are available to meet specific application requirements.

Regardless of the microform selected, the Telefiche scanner converts the successively encountered picture elements of a given microimage to a digital signal which is compressed prior to transmission over wideband communication facilities. Letter-size document images are scanned at a resolution of 200 lines per inch horizontal and vertical. A choice of receiving terminals is available. For long-distance transmission, a digital facsimile transceiver is provided. It operates at a typical speed of 30 to 40 seconds per

174

Figure 6-7. The TDC VMT-2000 is a high-resolution video terminal for the display of document images or machine-readable data. It is designed for the remote transmission from a central file of documents or microimages. (Courtesy: Terminal Data Corporation)

letter-size page and produces hardcopy output. It also can be used as a conventional facsimile transceiver for the transmission of paper documents. For local document transmission, a high-speed printer operates at 8 to 10 seconds per letter-size page. It also can be optionally configured to operate as a computer line printer. For applications that do not invariably require hardcopy output, a unique dual display workstation is a microprocessor-controlled computer and communications terminal with two screens. One of the screens is essentially a "soft copy" facsimile receiver. Document images displayed on the screen can be recorded on paper, at the operator's discretion, by one of the printers described above. The second screen is a conventional alphanumeric display designed to access a computer-maintained index in the manner of the CAR systems described in Chapter Five. Planning Research Corporation has developed a computer-assisted retrieval module—called the Intelligent Query System—for this purpose.

In addition to the Telefiche System described above, several manufacturers have developed microimage transmission systems employing closed-circuit television and remote video monitors with adjacent hardcopy printers. Companies active in this area include Infodetics, Terminal Data Corporation,

TERA, and Ragen Information Systems. In some cases, the microimage transmission capability is an optional component within a computer-assisted retrieval system.

Video Technology and the Office

The use of closed-circuit television for microimage transmission indicates the close relationship between facsimile and video technology. Both technologies were developed at about the same time and, for a brief period in the 1940s, were viewed as potentially competing vehicles for the transmission of news information to homes—a competition in which television emerged as the clear winner. From the technological standpoint, both facsimile and television employ "raster scanning" in which information is transmitted as a succession of individual picture elements. Of course, television pictures are transmitted at a rapid rate designed to simulate motion, while facsimile is designed to transmit and record still pictures of documents. Television technology likewise has emphasized entertainment for the consumer market, while facsimile equipment—as previously discussed—is intended for business use. There has been, however, considerable recent interest in the application of video technology to the storage and transmission of office-generated information.

To facilitate interpersonal business communications, for example, Picturephone service has been available on a limited bases for a number of years.[11] As the name implies, Picturephone technology combines voice and video communications. A Picturephone set consists of a touchtone telephone, a microphone, a screen, and a video camera designed to transmit moving images. Videovoice, a similar product introduced by RCA Global Communications in the mid-1970s, differs from Picturephone in the transmission of a changing series of still, rather than moving, pictures. The obvious advantage of these services is that they permit face-to-face communication over long distance. As such, they constitute a potential alternative to travel in applications where voice communication alone is not sufficient. The historical disadvantage of video-augmented telephone services, however, has been cost. Picturephone service, where available, normally costs about 10 times as much as conventional telephone service. But managers are increasingly aware of the high cost of business travel in terms of transportation and lodging expenditures, as well as in diminished productivity resulting from time spent in transit. It is consequently possible that Picturephone and related services will play a more important role in office communications during the 1980s.

The effects of management's awareness of the high cost of travel likewise are being seen in the related technology of *videoconferencing*.[12] As is true of Picturephone communication, a videoconference combines voice and image transmission. But, unlike Picturephone with its person-to-person orientation, videoconferencing typically involves two or more groups of persons, each

Figure 6-8. Videoconferencing combines voice and image transmission technologies to facilitate communication between two or more groups, each meeting in specially designed, geographically separate studios. Multiple cameras and monitors can be used to transmit and receive images. (Courtesy: American Telephone and Telegraph)

group meeting in specially designed, geographically separate studios. The videoconference studio is equipped with voice-activated video cameras, one or more television monitors, microphones, speakers, and related support equipment. Special desk-top cameras may be used, for example, to transmit images of charts, graphs, drawings, or other documents relevant to a given presentation or discussion. Special terminals are available for the transmission of handwriting from one meeting place to another. Such handwritten communications can be recorded on paper or projected onto a wall-mounted screen for group viewing.

Alternatively, Bell Telephone Laboratories has developed an "electronic blackboard" which transmits a chalk-written message to a remote video monitor. As was true of Picturephone, the prior development of videoconferencing has been impeded by high installation and transmission costs. The availability of video transmission capabilities through satellite carrier is, however, likely to reduce communication costs. A number of large corporations are establishing videoconferencing studios, viewing the expense as a cost-effective alternative to the conventional practice of having managers or other participants travel to a predetermined meeting location. For smaller organizations with occasional videoconferencing requirements, studio facilities can be rented, on an as-needed basis, from American Telephone and Telegraph, Holiday Inns, and other organizations. The availability of such shared studio facilities is likely to increase in the coming years.

Turning from video-based information transmission to information storage and retrieval, one notes that there has been considerable recent discussion of the potential of video or optical disks in office applications. Chapter Three previously discussed the likely future role of optical disks as an economical, high-density medium for the online storage of machine-readable, computer-processible data in time-sharing applications. This discussion deals with the possible application of video technology to the office-based recording and storage of document images as well as the machine-readable data and text generated by word processors, intelligent terminals, and small business computers. In such applications, video disks are a frequently mentioned alternative to paper documents, microforms, and magnetic recording media. The discussion in this section is necessarily limited to the *potential* of video disk, since—at the time this chapter was written—no office-oriented video recording and storage systems were generally available for sale, although several prototype systems had been installed in government agencies. As of early 1981, practical applications of video disk technology were limited largely to home entertainment where initial equipment sales were reportedly disappointing. Several industrial applications, involving training materials, likewise have been reported.[13]

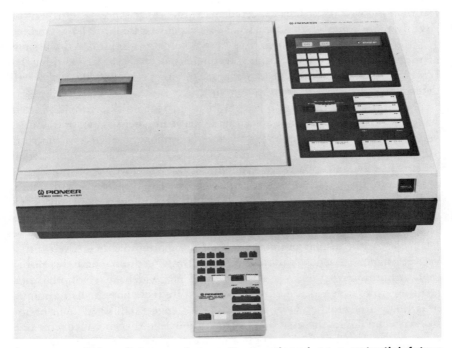

Figure 6-9. Video disks are frequently mentioned as a potential future alternative to paper documents, microforms and magnetic recording media. As of 1981, however, practical applications of video disk technology were limited largely to home entertainment where specially designed units use lasers or other mechanisms to play previously recorded disks. (Courtesy: U.S. Pioneer Electronics)

The several alternative video disk recording technologies have been described widely in newspapers and magazines. The first available system—developed jointly by Philips and MCA—uses a low-power laser to read information recorded in the form of microscopic holes in a plastic-coated aluminum disk. Image recording capabilities are not provided. The disks themselves must be prerecorded, typically in quantities sufficiently large to achieve a relatively low unit cost. In its present form then, video disk recording is a form of publishing in which information is produced in an edition of specified size. At the time this chapter was written, laser-based video disk systems were being sold in the United States by Magnavox and by Pioneer Electronics. The selection of available prerecorded disks consisted of approximately 100 titles, most of them motion pictures. Two alternative video disk recording systems—developed by RCA and Matsushita, respectively—have been announced for availability in 1981. Rather than using a laser, the RCA system will operate in a manner similar to conventional phonograph records, with a stylus travelling in grooves in the disk surface to detect information in the form of electrical capacitance signals. The Matsushita system likewise employs capacitance technology, but does not feature a grooved disk. It is expected that these capacitance systems will be slightly more compact and somewhat less expensive than the laser-oriented systems described above. There have been a number of informed discussions of the superiority of one recording technology over the other. Ultimately, however, the availability of a wide selection of prerecorded disks will determine the success of a given system.

From the standpoint of office automation, video disk technology offers several potentially significant advantages. Recording density is, for example, very high. A single platter about the size of a phonograph record can store 10^{10-11} bits, or somewhere between 50,000 and 70,000 pages of information. It is likewise an apparently economical storage medium, subject—at present—to the constraint of edition size discussed earlier. Furthermore, while magnetic and photographic recording technologies trade off access time against the cost per bit stored, video disks offer both rapid access and economical storage. Finally, video disk systems offer the familiarity of television for the display of information—an important factor in overcoming user resistance to a new information-processing medium—and the ability to transmit information, via satellites or other wideband communication facilities, to remote video monitors. But, despite these potential advantages, it is uncertain whether video disks can challenge established information storage technologies successfully.[14] Skeptics point out that the video recording systems developed for office applications during the late 1960s and early 1970s were unsuccessful. Those systems utilized video tape rather than disks, were very expensive compared to alternative technologies, and suffered problems of image quality and equipment reliability which will have to be resolved in future systems. Successful office applications of video disk will further require the development of a high-quality information recording capability designed for office operation. The economics of video disk recording likewise must change to eliminate the need to make multiple copies of a given disk to achieve a reasonable unit cost. Finally, the archival potential of video disk recording must be demonstrated if it is to compete favorably with micrographics and other established information storage technologies.

CHARACTER-ORIENTED MESSAGE TRANSMISSION

TWX and Telex

The character-oriented message transmission systems are so-called because they transmit information as a sequence of individually encoded characters rather than as a signal representing the light reflectance properties of successive picture elements in a pre-existing subject copy. With the exception of optical character recognition (OCR) technology, discussed later in this chapter, the character-oriented message transmission systems employ keyboard send-receive (KSR) or automatic send-receive (ASR) terminals for message preparation prior to transmission. In some cases, word processing equipment may be utilized.

TWX (Teletypewriter Exchange Service) and telex—the telegraph-derivative "record carriers"—are the historically dominant, and still most prevalent, forms of character-oriented message transmission. The two services are similar in operation. In the simplest point-to-point configuration, a message is key-entered at a typewriter-like terminal for transmission over a switched network of telegraph-grade lines to a designated compatible receiver that prints it onto paper. In most cases, the message is prepared on punched paper tape prior to transmission and is played back automatically through the terminal to minimize online time. Reception does not require an operator in attendance.

In the manner of all character-oriented transmission devices, the originating TWX or telex terminal converts the operator's keystrokes into a coded sequence of bits representing individual characters. The two services differ, however, in the type of terminal and coding scheme utilized, as well as in transmission speed, network coverage, and rate structure. TWX is a North American transmission network operated by Western Union. The most commonly encountered TWX terminal is a Teletype model 33 with a paper tape reader/punch. It features a typewriter-like four-row keyboard and transmits characters at 110 bits per second (10 characters per second), using the American Standard Code for Information Interchange discussed previously in Chapter Three as a widely used code in terminal-to-computer communication. Other ASCII terminals can be used for TWX transmission through an interface provided by Western Union, and several recently

Figure 6-10. While telex is a relatively old telecommunications technology, it continues to play an important role in office applications. Newer telex terminals are microprocessor-controlled and feature random-access memories for the simplified preparation and temporary storage of messages prior to transmission. (Courtesy: North American Philips)

introduced terminals have the ability to address either the TWX or telex networks. TWX rates vary with both time and distance on a scale ranging from 30 to 52 cents per minute for user-dialed transmissions within the 48 contiguous United States. Fractions of a minute are charged as whole minutes. Transmission costs between TWX terminals in the United States and Canada range between 31 to 90 cents per minute. These costs do not include prorated terminal equipment charges, network access charges, or the labor associated with message preparation and transmission. At the time this chapter was written, for example, the rental charge for a Teletype model 33 TWX terminal, including network access charges ($16.50) and maintenance, was $95.50 per month. At a volume of 500 messages per month (25 messages per day), the cost to transmit 125 words from New York to Los Angeles would be approximately $1.40 per message, assuming 7 minutes of operator labor at a wage of 6 dollars per hour.

Western Union also provides domestic telex service within North America. Unlike TWX, however, telex geographic coverage is international, and overseas transmissions from North American terminals must utilize one of the international telex carriers, such as French Telegraph Cable Company (FTCC), ITT World Communications, RCA Global Communications, TRT Telecommunications, or Western Union International—a division of Xerox Corporation not to be confused with Western Union, the domestic TWX and telex carrier. These carriers—which are located in the "gateway" cities of New York, Washington, San Francisco, New Orleans, and Miami—can be accessed through the domestic telex network or, in the "gateway" cities themselves, via a direct private wire connection. Where there is a high volume of overseas telex traffic, a leased international wire can be obtained. The most commonly encountered telex terminal, the Teletype model 32, features a three-row keyboard. A predefined sequence of shift keys is used to generate the required alphabetic, numeric, punctuation, and control characters. As an electromechanical device, the Teletype 32 is large and relatively noisy. Several vendors recently have introduced quieter desk-top telex terminals with conventional four-row keyboards. Customers can utilize telex-compatible terminals of their own selection. As is true of customer-provided TWX terminals, a network interface is required.

Within North America, telex terminals transmit characters at 50 bits per second (66 words per minute) using the Baudot code. International telex transmissions occur at the same speed, but utilize the CCITT No. 2 Code, which differs from the Baudot code in several minor respects. As is true of the TWX network, telex rates are based on a combination of elapsed time and distance, on a sliding scale ranging from 22.5 to 52.5 cents per minute for transmissions within the 48 contiguous United States. Unlike TWX, fractions of a minute are charged proportionately. But, because telex transmission speed is slower than that of TWX by approximately 34 words per minute, the cost to transmit certain messages may prove higher with TWX than with telex. In late 1980, the monthly rental charge for a telex terminal, including maintenance and network access charges ($30.00), was $96.50. At a volume of 500 messages per month, the cost to transmit 125 words from New York to Los Angeles is approximately $2.00, assuming 8 minutes of operator labor (message preparation and transmission) at a wage of $6.00 per hour. International telex charges vary with the record carrier selected.

The foregoing discussion assumes that TWX and telex subscribers will initiate transmission directly to a compatible terminal. While such point-to-point transmission has been the historically dominant form of TWX and telex communication, Western Union and the international telex carriers offer a number of special services which enhance the power and versatility of TWX and telex as vehicles for time-sensitive office communications. Current trends in transmission network design emphasize the use of computers as intermediaries that permit communication between otherwise incompatible terminals. This is the case, for example, with the ITT Fax-Pak network described in the preceding section on facsimile. As discussed above, the TWX and telex networks utilize different terminals, transmit messages at various speeds, and employ a number of coding schemes to represent individual characters. For domestic communications, the Western Union Info Master message-switching computer resolves these incompatibilities by performing the necessary speed and code translations to allow TWX terminals to transmit messages to telex terminals and vice versa. As might be expected, charges for Info Master transmissions are somewhat higher than for subscriber-dialed TWX or telex communications. TWX users also can utilize Info Master to access the international telex carriers, although most of the international carriers operate computer-controlled networks which will accept direct TWX communications. Graphnet, the North American value-added carrier mentioned in the preceding discussion of facsimile, is unique in accepting TWX and telex communications for retransmission to facsimile receivers where they are printed as paper messages.

TWX and telex users also can access the Western Union Info Master computer to transmit urgent messages via domestic telegram or international cablegram to persons or organizations who are not TWX or telex subscribers. Western Union delivers domestic telegrams within two hours by phone or five hours by messenger. The delivery time for international cablegrams varies with the rate. Where next-day message delivery is satisfactory, TWX and telex users can access the Info Master computer to send Mailgrams, as discussed later in this chapter.

As is true of telephone communications, a busy signal results if the terminal intended as the recipient of a TWX or telex communication is occupied. In point-to-point communication, the operator of the sending terminal must resubmit the message at some later time. However, Western Union Info Master and the computer-controlled international telex carriers do offer store-and-forward service in which such messages are held for later retransmission. The message originator receives a confirmation when delivery is accomplished. At the originator's option, messages that cannot be transmitted successfully via TWX or telex may be delivered via telegram or cablegram. Apart from retransmission necessitated by busy signals, store-and-forward capability can facilitate the transmission of messages involving changes in time zones. Western Union's Nite Cast service, for example, accepts TWX or telex messages during the day for delivery after 8 P.M. at a 30 percent discount. With some carriers, store-and-forward capability also provides an interface between faster communication devices and the relatively slow TWX and telex networks. As an example, the STORTEX service offered by TRT Telecommunications permits access to the international telex network from devices operating at speeds up to 1200 bits per second. The received message then is forwarded through the international telex network at the standard speed of 50 bits per second.

Western Union Info Master and the computer-controlled international telex carriers allow the TWX or telex terminal operator to transmit the same message to multiple recipients. With Info Master, for example, the same TWX or telex message can be sent to as many as 50 different addresses via any combination of TWX, telex, Mailgram, or telegram services. For applications in which TWX or telex messages are transmitted routinely and frequently to a predetermined list of recipients—such as managers of all regional sales offices—Western Union RediList service uses the Info Master computer to store and automatically address a list of up to 250 names.

Related Message Delivery Services

At the time this chapter was written, a proposal was under consideration for the establishment of an electronic mail delivery service to be operated by the U.S. Postal Service. The proposal was opposed strongly by various telecommunications companies, and its status was in doubt. The Postal Service is, however, a participant in Western Union's Mailgram Service in which customers submit messages to Western Union for electronic transmission to the participating post office nearest the addressee where they are printed and inserted in envelopes for delivery in the next business day's mail. Mailgram thus lacks the immediacy of TWX, telex, or facsimile, but it is faster than conventional mail delivery. Geographic coverage is limited to the United States and Canada. Customers can submit their Mailgram messages to Western Union in any of several ways. Occasional users can submit messages via telephone or in person. TWX or telex submission was mentioned in a preceding section. High-volume users can submit their Mailgram messages through their own computers, either electronically (via telecommunications) or physically (on magnetic tape). Western Union also makes available a special Mailgram display terminal, a text-editing device designed for applications in which text and address lists stored by Western Union can be merged with variable information, as required, to generate mass mailings for overnight delivery.

Motivated by the success of Mailgram, Tymnet recently introduced its Tyme-Gram service in which messages submitted electronically by customers working at conventional terminals are printed by the Tymnet computer and later taken to the post office for deposit in the mail. At the time this chapter was written, Tyme-Gram service was available in 30 cities within the United States.

Communicating Word Processors and OCR

Despite the capabilities outlined in the preceding sections, TWX, telex, and related message delivery services are limited in several important respects:

1. Messages submitted by originators in typewritten form must be rekeyed by the TWX or telex terminal operator. The terminal's keyboard is slower than that of a typewriter and, with regard to the telex terminal, requires more shifting operations.

2. While message preparation is usually performed offline prior to transmission, the punched paper tape that typically is used for message recording can be inconvenient to correct or otherwise edit.

3. The transmission speed of the TWX and telex networks is limited, as previously noted, to 110 and 50 bits per second, respectively.

4. The recipient of a TWX or telex message cannot capture it

conveniently in machine-readable form for computer storage, subsequent editing, or other manipulation.

5. The typography of the message is limited to upper-case characters and to a subset of commonly encountered punctuation marks and other symbols.

6. TWX and telex services are designed for relatively short, straightforward messages sent for information purposes only. TWX and telex are not suitable vehicles for the transmission, for example, of a multi-page report or for tabular presentations. Such messages can be transmitted, of course, via facsimile, but fax is not suitable for applications in which the intended recipient must revise or otherwise manipulate the message.

Communicating word processors can address many of these limitations successfully. As briefly discussed in Chapter Two, most automated text-editing systems can be optionally equipped with a combination of electronic circuitry and software which allows them to transmit data to, and receive data from, a computer or a compatible text-editing system. Some communicating word processors also can be configured for TWX and/or telex communications. When communicating with a remote computer, the word processor emulates either an ASCII or EBCDIC terminal and, as such, functions as the equivalent of the devices previously discussed in Chapter Three. This discussion is concerned with the use of word processors in point-to-point communications—that is, from one word processor to another. Compared to conventional TWX and telex transmission, the use of communicating word processors can eliminate the rekeying of typewritten messages generated originally by word processing equipment. The correction of errors and revision of messages is easily accomplished, and the full range of text production capabilities can be applied to message preparation. Thus, a multi-recipient message can be transmitted electronically to all or selected terminals from a prestored list. Prior to transmission, messages can be personalized by merging static text with variable information. The recipient can capture the message on magnetic media for further manipulation or retransmission. The typography of the message includes lower-case characters and, depending on the output mechanism utilized, special symbols. The speed of transmission varies with the communication facility selected. Over voice-grade telephone lines, for example, transmission typically occurs at a speed of 300 or 1200 bits per second, while conditioned leased lines will permit faster transmission.

The relationship between optical character recognition and word processing was discussed in Chapter Two. Most of the OCR readers designed for word processing applications also can be used for message transmission. Such machines offer the advantage of accepting input prepared on typewriters, thus permitting decentralization of the message preparation activity. In most cases, special typing elements must be used and specific formatting instructions followed, but productivity is typically through the elimination of repetitive keystroking at TWX, telex, or other special message transmission terminals. The exact mode of message transmission varies from application to application. In some cases, the OCR reader transmits directly to a remote printer where paper copies are reproduced. Alternatively, the OCR reader can transmit to a CRT terminal. OCR readers likewise can be configured to address the TWX and telex networks. In a commonly encountered configuration, the OCR reader operates online to a CRT terminal equipped with a paper tape punch. Characters in specially typed messages are recognized by the OCR reader and are transmitted to the terminal where they are displayed on the screen for proofreading and correction by an operator, as required. The corrected message then is recorded, character by character, on a TWX- or telex-compatible paper tape. The tape is taken to a TWX or telex terminal for transmission to a compatible receiver.

In the context of this discussion of character-oriented message transmission technologies, it is important to emphasize the distinction between optical character recognition and facsimile transmission. While both use optical scanning to determine variations in patterns of light reflected from a given document, facsimile technology does not attempt to identify individual characters. It merely transmits information about the tonal values or light-reflectance characteristics of successively encountered picture elements. OCR readers, on the other hand, attempt to identify individual characters based on internally stored definitions of the manner in which given characters reflect light. It transmits a signal in which individual characters are represented by their ASCII, EBCDIC, or other coded equivalents. The distinction between these seemingly similar technologies reflects the essential difference between image-oriented message transmission and character-oriented message transmission.

Computer-Based Message Systems

The TWX, telex, and other electronic mail and message technologies discussed in the preceding sections are designed to deliver a message to a designated terminal, either at the intended recipient's physical location or, with regard to Mailgram Service, to a nearby post office. They are used for messages that:

1. Are time-sensitive, in the sense that their impact or value would be lost if subject to the delays inherent in conventional physical delivery;

2. Require a written record, thus making the use of voice communication inappropriate; and

3. Are typically declarative in nature—that is, they do not require interaction or immediate recipient response. Many TWX and telex messages, for example, consist of order placements or notifications of the impending shipment of ordered materials.

While computers are used in such systems, their role is limited largely to the routing of messages between terminals. With the exception of the

Stored Mailgram or RediList services discussed above, the computer does not participate in message preparation or in the subsequent storage or retrieval of messages.

The systems described in this section differ from the previously discussed message delivery services in several important respects:

1. The computer's role is not limited to message switching, but extends to message preparation, storage, and retrieval;

2. Messages are transmitted to a specified person rather than to a specific terminal for later redelivery to a certain person. The computer serves as an intermediary which accepts messages, notifies designated recipients, and stores the messages temporarily or permanently, as directed. Recipients can receive messages at any compatible terminal.

3. The justification for such services are different than those employed with TWX, telex, or other message delivery services. They are used less for the transmission of time-sensitive communications than as tools for the enhancement of managerial productivity in a manner discussed later in this section.

One of the earliest and best-known examples of a Computer-Based Message System—ARPANET—was developed to facilitate communication among members of the research community within the United States Department of Defense. It has been followed by a number of individually developed systems designed to facilitate communication between managers in both the public and private sectors. Citicorp, for example, developed such a system to facilitate communication between supervisors and subordinates in the mid-1970s. Similarly, many universities have developed computer-based message systems for communications among faculty, students, and computing center personnel. By the late 1970s, a number of systems were publicly available—either through computer service bureaus on a time-sharing basis or in the form of prewritten software for implementation on a customer's own computer. Several examples were mentioned earlier in this chapter. In 1980, as previously noted, both Telenet and Tymnet announced the inauguration of Computer-Based Message System services, thus tying electronic mail and message systems even more closely to the telecommunications industry. In an interesting recent innovation, companies such as ROLM Telecommunications have implemented computer-based message systems within a telephone system controlled by a digital PABX. Managers, equipped with special workstations consisting of a terminal and a telephone, thus are given the option of communicating by voice or via a key-entered message.[15]

While the available systems differ in their specific implementations, they are remarkably similar in concept.[16] They typically consist of an integrated series of programs that permit the preparation, transmission, reception, and disposition of messages. The input device can be either a display terminal or a teleprinter of the types previously described in Chapter

Figure 6-11. The ROLM CBX Electronic Message System™ is one of an emerging group of office machines designed to give the user the option of communicating by voice or via key-entered message. Linked to a computerized branch exchange, it includes a telephone handset, a CRT with keyboard, and a hardcopy printer. (Courtesy: ROLM Telecommunications)

Three. The operator logs on and directs the computer's operating system to make the Message Preparation program available. The program is loaded and prompts the operator for the addressee of the message. While earlier systems used special codes to identify the intended recipient of a message, most newer ones allow addressing by the recipient's name. The message preparation program then prompts the operator for the subject of the message. This corresponds to the subject line that typically appears in the heading area of a typewritten memorandum. The inclusion of a distinct subject line permits mail scanning, a convenient feature described below. Following the entry of the subject line, the program prompts the operator for the text of the message, which it accepts passively without regard to its format or content. Alphanumeric characters, punctuation symbols, carriage returns, and line spacing all become part of the message as entered. Some systems provide special commands which permit the insertion of data from computer-maintained files into the message. To facilitate the correction of typographical errors or to permit modifications in entered text, most newer systems provide access to a text-editing program as part of the message preparation subsystem. As previously discussed in Chapter Two, these computer-based programs are somewhat inconvenient in that they rely heavily on key-entered codes and commands to perform tasks which, on word processing equipment, would be accomplished through special dedicated keys. Of course, the computer-based message system user can utilize a communicating word processor, or an intelligent terminal with text-editing capabilities, for message preparation and input.

Once the entry of the text is completed, the operator appends a command that initiates transmission. Most systems allow the same message to be sent to multiple recipients as separate communications, carbon copies, or blind carbons. The originator of the message can further request either acknowledgement of receipt or an immediate reply. Some systems even allow the originator to specify transmission at a specified date and time. The computer stores the transmitted message on a disk or other direct access storage device which can be viewed as the designated recipient's "electronic mailbox." When the designated recipient next signs on to use the computer, for whatever reason, he or she typically will be notified of the existence of any transmitted messages. Specific output options vary from system to system. In most cases, a scanning feature generates a list consisting of the originator, date, time, and subject line of each received message. The recipient then can display the full text of selected messages, if desired. Alternatively, the reading of received messages can be deferred until some later time.

Message disposition features vary from system to system. The most primitive computer-based message systems delete messages once they are read, forcing the recipient to print a copy for later reference. Others allow the recipient to return the message to online storage for later rereading. Newer systems, however, provide an increasingly broad range of storage and retrieval alternatives. If the content of the message is a draft of a report, for example, the recipient can edit it and return the edited version to the originator. Most systems allow the recipient to forward a message electronically, with or without comments, to designated persons. Messages can be relegated to named files and, in the most sophisticated systems, keywords or other index terms can be assigned to facilitate later retrieval by a combination of subjects, correspondent name, date, or other parameters. Alternatively, the message heading or the entire text may be scanned for the occurrence of a specified word or phrase. Originators can process, store, and subsequently retrieve their outgoing messages in the same manner.

The cost of computer-based message systems depends on the particular hardware and software configuration employed. The unit cost per message typically declines with volume. With regard to most commercially available systems, the cost per message compares favorably with that of TWX or telex communications. As previously noted, the justification for computer-based message systems frequently is based on improvements in the productivity

```
ELECTRONIC MESSAGE SYSTEM   2.28.80, 4:20 PM
SEND OR RECEIVE
? send
TO:
? N Van Coevering
SUBJECT:
? Telecommunications Planning Committee Meeting
TEXT:
? The Telecommunications Planning Committee will consider your
? proposal at its meeting on March 5.  It would be extremely helpful
? if you would send me 5 additional copies of your cost estimates
? as soon as possible.  Thanks.
? .send
MESSAGE SENT   2.28.80, 4:25 PM
DO YOU WISH TO CONTINUE?
```

Figure 6-12. Computer-based message systems are an alternative to conventional voice communications in applications where interaction is not required. In the above, typical example, the computer prompts the user for the entry of the recipient's name, subject, and text of the message. The date and originator's name are derived automatically. The most advanced systems permit the sending of the same message to multiple recipients and the filing of transmitted or received messages.

of managerial personnel. As briefly noted in Chapter One, much management information comes through informal communications with peers and subordinates. These communications take the form of memoranda, face-to-face discussions, meetings, and telephone calls. Computer-based message systems, in effect, extend the concept of a computer-based management information system, as discussed in Chapter Three, to the informal communications among individuals.

Computer-based message systems are further designed to improve the efficiency of managerial communications. Studies confirm, for example, that most managers spend substantial portions of their time making telephone calls. A high percentage of all attempted telephone calls—perhaps as high as 70 percent in some instances—are unsuccessful because the called party is either absent or otherwise unavailable.[17] Many phone calls involve a simple message and do not require immediate interaction with the called party. Such phone calls could be effectively replaced by computer-delivered messages, with a consequent savings in time and elimination of needless interruption. The goal of computer-based message systems is to minimize time-wasting activities and to enable managers to maintain sustained concentration on a given work task, thus improving productivity. While a ringing telephone demands to be answered, computer-based message systems allow a user to determine both the time when communications will be received and the priority to be accorded to them. Computer-based message systems are likewise flexible in permitting access from any terminal anywhere. They are thus well suited to applications requiring communication between travelling managers and peers or subordinates. A further justification for such systems involves the improvement of communication between supervisors and subordinates, thus permitting a given supervisor's span of control to be extended by enabling him or her to communicate more effectively with a larger number of subordinates. This justification, which has been most closely associated with the Citibank experiments with electronic message systems, seeks to decrease office costs by reducing the number of high-paid managers required to supervise a given operation.[18]

In addition to advantages designed to improve managerial productivity, it is important to note the impact of computer-based message systems on clerical personnel. A reduction in the number of telephone calls resulting from the transmission of written messages can be translated, for example, into corresponding reductions in the number of required administrative secretaries or other clerical support personnel. Similarly, freed of telephone interruptions, clerical personnel can perform other work more effectively. Perhaps the greatest area of potential clerical labor savings, however, is derived from the automatic storage and retrieval of messages provided by newer computer-based message systems. In conventional message handling systems, the duplication, routing, and filing of internally generated memoranda accounts for a substantial portion of office costs. The implementation of a computer-based message system makes many of those activities unnecessary. Likewise, elimination of a substantial portion of internal memoranda can lead to a reduction in the size of internal mailroom operations.

Computer Teleconferencing

As used in this chapter, the term *teleconferencing* denotes a special variant of computer-based message system in which three or more persons, working on the same project or with similar subject interests, communicate with one another via terminals. Thus, teleconferencing differs from videoconferencing which, as described earlier, relies on real-time audio-visual presentation rather than nonconcurrent key-entered communications.

There are several types of teleconferencing. In the most simplistic type, a project leader or other person sends a message to multiple recipients for comment. Individual responses are appended to the original message and can be reviewed by the other participants. This, in turn, may provoke responses that are themselves made available for review. The result is a recorded dialogue which can be reviewed, in part or in its entirety, by all authorized persons. It is not difficult to see the advantage of such teleconferencing in facilitating communication among: (1) researchers, scientists, or engineers working on projects; (2) journal editors and reviewers of scholarly papers; (3) committee members working on a draft of a position paper; or (4) other groups of geographically scattered individuals working on a common endeavor. As is true of conventional computer-based message systems, teleconferencing removes barriers of time and geography. It allows individuals to participate at their own pace, without the interruption and contention that is too often characteristic of conventional meetings or conference telephone calls. As a further advantage, a permanent transcript of all discussions relevant to a given project is created. That transcript, in its entirety or in part, can be stored for later retrieval.[19]

7
Integrating and Implementing Automated Office Systems

Integrated information systems • Some nontechnological considerations

The preceding chapters discussed word processing, computers, micrographics, reprographics, and electronic mail as more or less isolated technologies, emphasizing the concepts and terminology essential to an understanding of the current state of the art in office information processing. As users gain increased experience with automated office systems, however, considerable interest has developed in potential interrelationships between technologies. Similarly, experience with automated office systems has led to an awareness of the social and behavioral constraints which limit the application of technology. These matters, largely neglected in the preceding discussion, are treated briefly in this concluding chapter.

INTEGRATED INFORMATION SYSTEMS

Basic Concepts

For purposes of this discussion, an integrated information system is one that utilizes some combination of data, word, and image processing technologies to accomplish a given office-related information-processing task.[1] The motive for such systems is the same as that for any type of rationalized information processing: improved cost-effectiveness through either cost reduction or increased productivity. The current interest in integrated information systems is based on the availability of improved technological tools and on changing attitudes toward information processing. Of course, the improved technological tools were discussed in the preceding chapters. Two previously described technological developments warrant re-emphasis, however, in the context of integrated information systems:

1. From the standpoint of equipment design, the most important

development of the 1970s was large-scale integration—the improvement in the manufacturing of electronic components that made possible the production of low-cost microprocessors and memory circuits. These computer-like components have been widely incorporated in a variety of machines, permitting the development of such previously unavailable products as microcomputers, "intelligent copiers," intelligent microfilmers, and high-speed digital facsimile transceivers. For the first time, there is a common "design bond" among various types of information-processing equipment. Data, word, and image processing tasks are being performed by increasingly similar machines. In addition, because they derive their capabilities from internally stored programs rather than from fixed circuitry, these new devices offer much greater power, flexibility, and adaptability than previous generations of equipment.

2. As described in Chapter Two, changes in the existing communications infrastructure will increasingly facilitate the transmission of information between different devices. Several value-added networks, for example, permit communication between incompatible facsimile transceivers or between facsimile transceivers and character-oriented terminals. Similarly, available and projected satellite networks are designed to transmit a mix of voice, character, and image-oriented messages. Ethernet®—the coaxial cable-based local network developed by Xerox in conjunction with Digital Equipment Corporation and Intel—is designed specifically to interconnect different types of information-processing equipment, and it is expected that such interconnections will increasingly be a feature of future communication systems.

While the concept of integrated information systems is not new—CAR systems combining computer processing with micrographic storage of document images, for example, date from the 1960s—the idea of integrating several technologies attracted comparatively little interest through the mid-1970s, largely because of the presumed superiority of computers as the sole state-of-the-art approach to information processing. Most systems analysts believed that those problems that were not readily solvable by computerization alone would eventually prove amenable to computer solution, given the rapid development of electronic technology. Alternative technologies consequently were viewed as transitional. One of the most important motives for integrated information systems stems from the failure of conventional data processing alone to solve several of the most important problems once viewed as logical candidates for computerization. As an example, even though the cost of online machine-readable storage has dropped continuously, as discussed in Chapter Three, the cost of converting information to machine-readable form remains very high. Similarly, there has been little progress in the development of machine-readable storage media possessing archival properties, while attempts to develop large-scale time-sharing word processing systems have proven unsuccessful from the standpoint of human engineering. As a result, many systems analysts now have a broader perspective on information-processing activities than their counterparts did 20 or even 10 years ago. They view a multiplicity of approaches as viable and recognize that a given problem might best be solved by employing a combination of technologies, each designed to address a specific facet of the problem. In this sense, they are truly "information systems" analysts as opposed to merely computer systems analysts, word processing systems analysts, micrographics systems analysts, or reprographic systems analysts, each of whom utilize one approach to problem solution.

But an integrated systems approach presents problems as well as advantages. The systems analyst, for example, must be familiar with both the fundamentals and current state of the art in various information processing disciplines. This difficult task is complicated by the fact that terminology varies from technology to technology, as does the structure of the industry that develops products and services. Specification writing can prove to be extremely difficult, and the analyst often faces procurement problems involving multiple sources in different industries. It also may prove impossible to clearly pinpoint responsibility for interface problems or system malfunctions. Yet, in many applications, the greater effort expended to develop an integrated solution to a given problem can improve the cost-effectiveness of a given system significantly. The remainder of this section discusses the prevailing approaches to the integration of automated office technologies, indicating the extent to which the various technologies discussed in preceding chapters can be readily integrated.

Multifunction Systems

The integration of automated office technologies can be accomplished in either of two broad ways:

1. By purchasing one of the preconfigured systems which combine the capabilities of several technologies in a single, multifunctional piece of equipment or system; or

2. By integrating several separate pieces of equipment into a system configured expressly for a given application.

The multifunctional approach, in which several technologies are pre-integrated by the manufacturer of a given system, now is employed on a limited basis by various vendors and is expected to increase in popularity. Intelligent terminals, which combine data and word processing with voice and message transmission capabilities, now exist in prototype. One such device, an experimental workstation developed by Siemens' Munich Research Center, reportedly permits the input of text and graphics via a combination of keyboard, light pen, and image digitizer, and is capable of either image- or character-oriented transmission and printing.[2] Similarly, several recent research studies forecast the future development of multipurpose terminals which combine a microprocessor/controller, a CRT or flat panel display with dedicated special function keys, a facsimile-type xerographic printer for either image- or character-oriented data, and local storage in the form of magnetic or optical disks. These devices will be furnished with prewritten word and data processing software and will support user programming in BASIC or other higher level languages. They will be capable of communicating with remote computers, TWX and telex machines, word processors, and facsimile transceivers.[3]

While some of these combined capabilities are available now, most existing multifunction equipment is of more modest scope. Chapters Two and Three, for example, noted the virtually identical hardware configurations employed by stand-alone word processing systems and microcomputers. Chapter Two further indicated that some word processing systems are incorporating data processing capabilities, including arithmetic and sorting functions, while several manufacturers of word processing systems offer financial management software and a BASIC interpreter. Similarly, as indicated in Chapter Three, many microcomputer vendors offer a word processing software package. Regardless of the method of implementation, a combined word and data processing system, as previously discussed, can be optionally equipped with a modem, thus adding the ability to communicate with remote computers in order, for example, to supplement local storage, retrieve information from centralized data bases, or access computer-based message systems. A few systems also can address the TWX or telex networks.

While word processing and small business computer systems can acquire multifunctionality through software enhancements, the most ambitious and sophisticated examples of the multifunctional approach to integrated information systems are the preconfigured office automation systems developed by companies such as Prime Computer, Datapoint, and Basic Four. Such systems typically employ a fairly powerful minicomputer as their central processor and feature faster printers and more extensive online storage capabilities than microcomputer-based systems. Depending on the system, 64 or more display terminals may be supported simultaneously. The terminals themselves are typically microprocessor-controlled and feature specially designed keyboards created to simplify utilization in both word and data processing applications. From the software standpoint, these office automation systems feature the conventional word processing capabilities described in Chapter Two, as well as advanced text management functions such as proofreading, automatic hyphenation, and word-for-word translation made possible by a powerful central processor and extensive online storage. The automatic filing and retrieval of processed text also may be supported, thus permitting the substitution of machine-readable records for paper documents. In terms of data processing capabilities, these systems are user-programmable in BASIC or other higher level languages. Some systems also provide a data base management capability. These systems also include

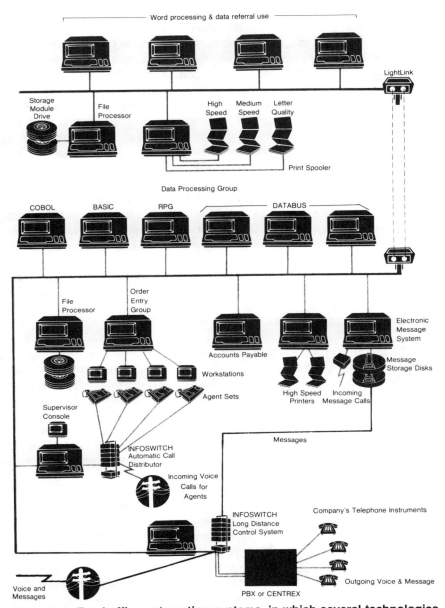

Figure 7-1. Total office automation systems, in which several technologies are pre-integrated by the manufacturer of a given system, are expected to increase in popularity. The Datapoint Integrated Electronic Office, for example, consists of user-selected combinations of computers, terminals and telecommunications components designed to perform data processing, word processing, message transmission, and related information processing tasks. (Courtesy: Datapoint Corporation)

software which permits the transmission of messages to other users or terminals. Some systems also will communicate with other computers or will direct messages to designated TWX or telex terminals. In some cases, the system may function as a digital PABX for the control of voice communications.

In addition to the word, data, and message processing capabilities described above, some of these office automation systems offer support for certain commonly encountered management activities. For example, they may provide automatic calendar maintenance to assist managers and secretaries in the scheduling of appointments and meetings—a potentially time-consuming task when performed manually, especially if the calendars of several persons must be coordinated. Calendar software maintains the personal appointment books of system participants. On instructions entered at one of the system's terminals, the calendars of specified persons can be examined to determine their availability for a meeting on a specified date. Alternatively, the system automatically will search the calendars of specified persons to determine those dates that are available for a meeting. Some calendar maintenance systems also include a "tickler" or reminder capability to inform users at a specified time that a previously entered task is to be performed or some previously specified action is to be taken.

Interfacing Automated Office Systems

Apart from preconfigured multifunction office automation systems, integrated information systems can be implemented by combining or interfacing two or more stand-alone systems, each embodying different technologies. Examples of such interfaces were described in preceding chapters. This discussion briefly will review the material presented in those chapters and consider other aspects of the current state of the art in interfacing technologies, indicating the extent to which various available systems can be combined.

As the point of capture for much office-generated information, considerable attention has been given to the potential interrelationship of word processing equipment and other automated office systems. Although several vendors recently have integrated dictation systems with telephone answering equipment, voice recording is, at present, a stand-alone technology. Sony recently announced a portable cassette-based text-editing typewriter that also can function as a dictation machine, but the true integration of dictation systems and text-editing technology will not occur until voice-input typing systems are developed. As indicated in Chapter Two, the development of reliable and versatile voice-input typing systems is unlikely in the first half of the 1980s. The use of digitized voice input for the transmission of "speech mail," however, may occur within the next several years. Such systems are reportedly under development.

Automated text-editing systems, the core of word processing technology, can be interfaced successfully with several types of automated office systems. Input prepared on a word processing system can be transmitted, for example, to a phototypesetter or photocomposer, thus eliminating much of the repetitive keystroking associated with the production of typeset text. Through the addition of optional communications capabilities, most word processing systems can be made to emulate a given online terminal, thus permitting the transmission of data to, or the reception of data from, a remote computer. When equipped with terminal emulation capabilities, a word processing system can draw on the computer's peripheral devices, notably more extensive online disk storage capacity and faster output peripherals. Thus high-volume document production jobs, which would prove time-consuming to print on the relatively slow typewriter-like output devices commonly provided with word processing systems, can be dispatched through a computer to a faster line printer or xerographic page printer. As

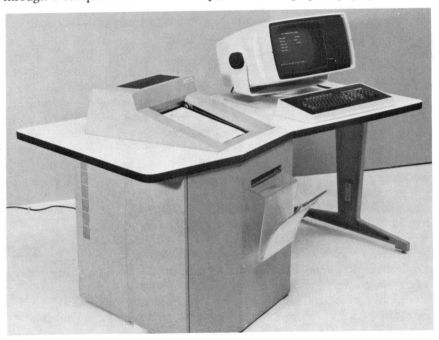

Figure 7-2. Some information systems analysts believe that the future implementation of integrated systems will be facilitated by the development of innovative, multi-function work stations which combine several technologies. In the recently introduced Ragen 1010 Information Management System, remote work stations feature a video terminal which can be used for the display of computer/user interaction or of document images transmitted from a remote micrographics data base. Images can be dispatched from the screen to an adjacent hardcopy printer. (Courtesy: Ragen Information Systems.)

noted in Chapter Five, however, the Xerox 9700 can interface directly with certain word processing systems, eliminating the need for the computer as intermediary.

Some recent interest likewise has been expressed in possible interfaces of word processing and computer-output-microfilm technologies, thus permitting the direct reduction of correspondence and other office documents to microform and also eliminating the need to establish, and later microfilm, paper files. The implementation of such an interface poses, however, several relatively complex problems. Each word processor outputs data as a serial stream of encoded characters with embedded machine commands designed to drive the various printers or other peripheral components in a word processing system. As an output device, a COM recorder responds to a combination of encoded characters and embedded machine commands, but the coded characters and commands generated by a given word processor are usually incompatible with those required by a given COM recorder. Not only may different codes be used to represent a given character, but the character generation capabilities of most word processors typically exceed the output repertoire of available COM recorders. Thus, while they can increasingly generate lower- as well as upper-case characters, most COM recorders are incapable of underlining, printing superscripts or subscripts, or generating special scientific symbols or accented characters. In addition, a coded command that initiates a given output action on a printer local to a given word processor may cause a COM recorder to initiate a different output action. As a further complication, most COM recorders cannot respond to commands for such special output actions as automatic centering and decimal alignment. Thus, translation software must be developed to convert the output of a given word processor to a COM-compatible format.

In the late 1970s, the 3M Company demonstrated an interface which enabled its display-oriented word processing system to generate a magnetic tape formatted for output on one of the 3M COM recorders. At that time, the product generated too little interest to warrant further development. While alternative approaches involving, for example, floppy disks have been suggested, no direct interfaces between word processing systems and COM recorders are currently available. Users interested in interfacing their word processors with COM recorders consequently must utilize an indirect approach in which word processor output, consisting of encoded characters and embedded commands, as previously described, is transmitted to a remote computer for translation to a COM-compatible format by custom-developed software.[4]

If the present interfaces between micrographics and word processing appear tenuous, the interrelationship of micrographics and other automated office technologies is better understood. COM and CAR technologies, both discussed in Chapter Three, demonstrate the well-established interface between micrographics and computers. Similarly, the relationship between micrographics and the broader field of reprographics is reflected in the large number of available reader-printers and enlarger-printers capable of creating full-size paper copies from microimages. Chapter Four likewise mentioned the existence of the COMp 80/2, a powerful COM recorder with output capabilities similar to those of photocomposition devices. The most widely publicized interface between micrographics and electronic mail and message systems—microfacsimile—was discussed in Chapter Six. While it provides for the direct transmission of images recorded on film, a similar effect can be achieved indirectly by first using a reader-printer to make paper enlargements of microimages then transmitting those enlargements to remote locations via conventional facsimile. Microforms likewise can be used as the storage medium for communications transmitted and received by computer-based message systems.

The importance of interfaces between other automated office technologies and electronic mail and message systems will increase as electronic transmission replaces physical message delivery in a greater percentage of business applications. As discussed in Chapters Two and Six, word processors equipped with optional communications capability can be used to prepare input to computer-based message systems. Similarly, communicating word processors can be used for point-to-point message transmission where the receiving device may be another communicating word processor, a TWX terminal, or a telex terminal.

The integrative role of two devices—OCR readers and "intelligent copiers"—has been discussed in preceding chapters, but is sufficiently significant to warrant repetition. By virtue of its ability to interface with word processing, data processing, and electronic mail and message systems, optical character recognition can serve as a bridge between otherwise incompatible technologies. If, for example, text generated by a word processor has been recorded in human-readable form on COM-generated film, that same text can be converted back to machine-readable form for subsequent editing by first using a reader-printer or enlarger-printer to create paper copies for input to an OCR reader operating online to a word processor.[5] Similarly, OCR can provide an interface between incompatible word processing systems or between a given word processor and a TWX or telex terminal. Given their ability to reproduce information recorded in either human-readable paper or machine-readable magnetic form, "intelligent copiers" often are considered a convenient focal point for the integration of technologies. Available devices already can accept input from word processors and computers, generate multiple type fonts in the manner of phototypesetting equipment, and transmit character-coded information electronically. Given the likely future incorporation of facsimile transmission capabilities, "intelligent copiers" readily will permit the interfacing of word processing, data processing, micrographics, reprographics, and electronic mail and message systems.

SOME NONTECHNOLOGICAL CONSIDERATIONS

Documenting Automated Office Systems

In keeping with the primary focus of this book, the preceding chapters have dealt exclusively with the technological foundations of office automation. While a thorough understanding of word, data, and image processing technologies is essential to the office systems analyst, an automated office system is comprised of equipment and people. The successful automation of office activities requires a systematic approach to the organization and functioning of the office's human resources. From the administrative standpoint, documentation is the embodiment of that systematic approach.

Documentation can be broadly defined as recorded information pertinent to the development of a system or the performance of an activity. Although the term carries the connotation of a paper record, documentation may be recorded on microfilm or on magnetic media generated by word processors or computers. It is expected that the utilization of these alternative forms of documentation will increase in office applications. Regardless of form, the obvious purpose of documentation is to insure that the characteristics of a given system will be understood by those persons who have a need to know about the system. The careful documenting of automated office systems is important for a number of reasons:

1. The implementation of automated office systems is a complex activity which is often undertaken by a team of professionals with expertise in information processing. Documentation identifies the responsibilities and reports the actions of persons responsible for the design and implementation of a given facet of office automation. Good documentation facilitates communication among team members. Complete documentation can, in addition, simplify the later reactivation of an automated office project that has been terminated prematurely by eliminating much repetition of work.

2. Documentation is critical to the supervision and control of automated office systems. It enables management to assess the status of a given activity intelligently and, with regard to system design and development projects, it allows staff members to make informed decisions about their progress.

3. Documentation contains information essential to the maintenance of a system once installed. In the information-processing environment, where turnover of analytical personnel is characteristically high, a clear, comprehensible record of a given system's purpose, requirements, and specifications is critical to later system modification.

4. Documentation is the primary vehicle for instructing users of an automated office system. In many cases, the anticipated benefits of an automated office system are not fully realized because of inadequate attention to such documentation as written procedures and user manuals. The integration of an automated system into the office work environment obviously involves more than the mere installation of equipment.

Given its importance, a documentation plan should be established at the inception of an activity related to the automation of office operations. Documentation requirements and responsibilities should be identified in terms of purpose, intended audience, approximate content, and projected completion dates. Budgets for all automated office systems should include a line item for documentation. Office systems analysts likewise should develop a plan for the periodic review and updating of documentation. With regard to user manuals, for example, a program of evaluation and revision based on user recommendations should be provided.

In the data processing field, with which office automation is closely allied, documentation denotes the recording of information pertinent to a given computer application. The term is applied most often to software development, but a broader view extends its scope to hardware and data as well. Data processing documentation typically is divided into three categories, each of which has an office automation counterpart:

1. *Analytical* documentation, created at the system study and evaluation stage, describes an existing office system in terms of its advantages and disadvantages, presenting a statement of recommendations and functional requirements for the development of an alternative system. In automated office applications, the most commonly encountered example of analytical documentation is the feasibility study which customarily precedes the implementation of a given system.

2. *Development* documentation records the evaluation of alternative approaches to the design of a new system. It presents the rationale for the new system and describes it in detail, using narrative and graphic techniques, as required. Examples of development documentation include hardware and software specifications, procurement documents, and product evaluations.

3. *Operational* documentation, which presents information about the manner in which a given system is to be installed and operated, is critical to the success of office automation. Documentation in this category typically takes the form of manuals intended for two audiences: equipment operators and system users. In some applications, as in the case of online terminals or office-based computer systems, the user and the operator are the same person. In other applications, such as word processing, the user and operator are usually different persons with distinct information requirements. The word processing system user or word originator, for example, needs to know how to prepare work for submission to the word processing activity, while word processing operators need an understanding of equipment and work completion procedures.

While automated office documentation follows the familiar pattern established in data processing, there are some distinctive characteristics of office operations which have considerable significance for documentation practices:

1. There is much less emphasis on software in automated office systems.

Much equipment, for example, is preprogrammed by the manufacturer to perform a specified range of tasks, and software development is not the user's responsibility. This is the case, as previously described, with regard to word processing systems. Even in microcomputer systems, where programming is possible, much software is purchased from external sources rather than custom-developed. There is a third category of automated office equipment, represented by micrographics and facsimile systems, which have no software in the conventional sense of the term.

2. While data processing documentation is concerned primarily or exclusively with computers, automated office documentation must deal with a much broader range of technologies. While the primary orientation of automated office systems is electronic, some systems have a photographic component. As explained earlier in this chapter, some automated office systems intermix several different technologies, each with its own terminology and application characteristics. The person preparing automated office documentation consequently must have a background that is sufficiently broad to comprehend the varied attributes of the equipment utilized in a given system.

3. Automated office systems typically are implemented by diverse groups within an organization. A micrographics group, for example, may be responsible for the implementation of COM and CAR systems; an administrative services group may be responsible for the development of word processing systems; and a computer group may be responsible for the implementation of distributed data processing systems. In instances involving a mix of technologies, responsibility for implementation may be fragmented among several different groups. It is consequently difficult to establish and enforce documentation standards and practices.

4. Some automated office systems are not implemented in the classic data processing manner, which consists of a formal systems analysis followed by the development of an alternative system and subsequent approval and implementation of the alternative. In some cases, no formal analysis of the existing manual system is completed. Some manual office systems, in fact, prove very difficult to analyze in a formal fashion.

5. Many automated office systems retain a manual component which is essential to the operation of the automated system. In word processing applications, for example, realization of cost-effectiveness typically depends on routinization of the manual activities performed by administrative secretaries. As a result, much automated office documentation must concentrate on activities that may not be automated at all.

6. Data processing personnel have a common background, typically in the field of computer science. While they may not always take the time to document their work properly, they are at least aware of its significance, having been taught its importance as part of their formal training. Automated office systems are developed, however, by persons with varied educational

backgrounds. Many of them are unaware of the significance of documentation or of the desirable attributes that characterize effective documentation.

7. Persons responsible for the implementation of automated office systems typically have relied heavily on vendors for training and related user support. This reliance on external persons has tended to diminish the perceived need for custom-written user documentation. Recently, however, some vendors—in an effort to lower the cost of their products—have sharply reduced the extent of the commitment to user training and related support. In the future, it is consequently likely that users will have to assume the responsibility for developing much of their own instructional documentation.

Social and Behavioral Factors

It should be obvious from the preceding chapters that there are few remaining technological impediments to the automation of most office activities. In an increasing number of applications, system design and equipment procurement are, if anything, complicated by the existence of too many, rather than too few, satisfactory alternatives. While there are remaining areas—notably, computer software—which require further development, new technological approaches to office work are being introduced and refined faster than existing products can be utilized effectively. The acceptance of office automation as a means of realizing the productivity improvements discussed in Chapter One is constrained, however, by a complex mix of social and behavioral factors, each of which could be treated at book length. While a complete analysis is beyond the scope of this chapter, the purpose of this discussion is to provide a brief introduction to those social and behavioral considerations which influence the successful implementation of office automation.[6]

The problem of changing deeply engrained managerial work habits is a case in point. While decades of office automation have contributed to the increased routinization of clerical jobs, management activities remain highly unstructured. As discussed in Chapter One, this too often contributes to inefficient expenditures of time and effort. Rather than seeking productivity improvements by encouraging managers to persist in their existing work habits but try harder, office automation follows the classic pattern—which has proven so successful in agriculture and manufacturing—of using machines to change and enhance the way in which work is performed. Thus, instead of drafting correspondence in longhand for personal delivery to a nearby secretary, the manager will dictate to a remote recording machine for transcription by a centralized word processing facility; instead of frantically rummaging through desk drawers and file cabinets in search of a given paper document, the manager will sit at an online terminal, desk-top computer, or microform reader where vast quantities of information are made conveniently and quickly available; instead of wasting minutes or hours on unanswered telephone calls, the manager will route noninteractive communications through a computer-based message system. But, to enjoy the advantages of automated office operations, the manager must learn, for example, to prepare a succinct outline of his or her thoughts and express them clearly into a dictating machine. To utilize online terminals and computer-based message systems effectively, the manager must be able to type and feel comfortable doing so. To realize the benefits of online data bases and desk-top microform files, the manager must become accustomed to "softcopy" displays as an alternative to paper documents. In each of these instances, an educational reorientation is required. To facilitate such reorientation, there has been considerable recent discussion of the development of special terminals designed specifically for management, rather than clerical, personnel. It is difficult, however, to distinguish the few existing examples of such management terminals from those intended for clerical use, although it is likely that the concept will be articulated and improved.

Even more serious managerial adjustments are necessitated by the impact of automation on the office's social structure. The most frequently cited example of such impact is the elimination of personal secretaries—one of the traditional symbols of managerial status—in centralized word processing systems.

The potentially disruptive social effects of automation are likewise experienced by clerical personnel, many of whose jobs are altered radically by the introduction of systems that modify or eliminate such familiar office tasks as typing, filing, and telephone answering. As is true of managers, an educational reorientation is required of clerical personnel, but the psychological impact of automation on work performance should not be underestimated. In some offices, clerical personnel are able to view automation positively as a vehicle for job enrichment or career advancement. In others, however, the implementation of automated systems is accompanied by concern over skills obsolescence and eventual unemployment. To a large extent, this fear is justified, since one of the goals of automation is to reduce the quantity of labor required to yield a given quantity of output. The concern is not limited to clerical personnel, however, since—as noted in Chapter Six—some automated office systems are justified on the basis of potential reductions in the number of managerial and supervisory personnel. As was true in agriculture and manufacturing, the automation of office operations necessarily will be accompanied by reductions in certain segments of the workforce. Most economists contend, however, that automation has a beneficial long-term impact on employment. While temporary unemployment may be experienced in jobs that are directly affected by the introduction of office automation, the resulting improvements in productivity will stimulate economic growth which will, in turn, produce more jobs. Planning at the national level, however, will be required to insure that the labor force is retrained properly and is redirected into areas of future employment opportunity. In the office itself, there will continue to be a requirement for persons trained and skilled in the development, implementation, and operation of the automated systems discussed in preceding chapters.

GLOSSARY

A

acoustic coupler—a device that converts a terminal's electronic signal to an audible form suitable for transmission over telephone lines. An acoustic coupler is an alternative form of modem which features a set of rubber cups designed to cradle a telephone handset. Acoustic couplers are commonly provided on computer terminals and facsimile transceivers.

application software—programs that are designed to perform some user-specified task.

ASCII—American Standard Code for Information Interchange. An American National Standard binary-coding scheme consisting of 128 seven-bit patterns for printable characters and control of equipment functions.

automatic send-receive (ASR) terminal—a Keyboard Send-Receive terminal with peripheral attachments or memory circuits which permit the offline preparation of data prior to transmission or the capture and storage of incoming data.

B

basic—a programming language widely used in office-based small computer systems.

baud—a measure of the speed of transmission in information systems. Bauds measure the number of times that a given communication line's condition changes per second.

bit—a binary digit. The smallest element of binary machine language represented by a magnetized or optical spot on a recording surface. Six to eight bits are required to form a character or byte.

bits per second (BPS)—a measure of the speed of transmission in information systems.

blip—a small opaque mark recorded beneath all or selected microimages in computer-assisted microfilm retrieval systems. The blips are counted by readers or reader-printers equipped with appropriate electronic circuitry in order to advance the film to a specified number frame.

boilerplate—in word processing applications, prewritten pages, paragraphs, or text segments that are used repeatedly in document production.

bus—in computer systems, the network of connecting wires along which electrical signals travel among the central processing unit, memory, and peripheral devices.

byte—a sequence of adjacent binary digits that represent a character.

C

cassette—(1) A double-core container enclosing processed microfilm designed to be inserted into readers, reader-printers, and retrieval devices. (2) A lightproof container of rigid metal or plastic containing film for daylight loading in cameras. (3) A container for magnetic tape.

coaxial cable—a broadband communications facility that is expected to play an increasingly important role in intra-building transmission of data and images. The Xerox Ethernet® system is an example of a coaxial-based communication facility.

communicating word processor—a text-editing system equipped with electronic circuitry which enables it to transmit data to, or receive data from, a computer, another communicating word processor, or—in some cases—a TWX or telex machine.

computer-assisted make-up and imaging systems (CAMIS)—a term used to denote the increasing convergence of reprographic and computer technologies, thereby permitting new approaches to the reproduction and dissemination of information in paper form.

computer-assisted retrieval (CAR)—the capability to have micrographic images located or identified by commands initiated through a computer terminal.

computer-based message system (CBMS)—a form of electronic mail and message system in which a computer serves as the intermediary for the storage and routing of messages between individuals. In effect, a "mailbox" is established on disks or other direct access storage devices for each authorized user. Messages deposited in a given mailbox can be retrieved, printed, filed, or erased at the recipient's discretion.

computer-input microfilm (CIM)—the process of reading data contained on microfilm by a scanning device and transforming this data into a form suitable for computer use. Synonymous with CIM.

computer-output microfilm (COM)—microforms containing data produced by a recorder from computer-generated electrical signals. Synonymous with COM.

correcting typewriter—an IBM Selectric® or other single element typewriter which features a special adhesive-backed ribbon designed to remove previously typed characters from paper. The Correcting Selectric is used in some typewriter-based text-editing systems.

CRT Terminal—an electronic tube in which a well-defined and controllable beam of electrons is produced and directed to give a visible or otherwise detectable display or effect. The displayed image may be read visually, microfilmed, photographed, or recorded in some other manner.

D

daisywheel printer—in word processing and computer systems, a device which features an interchangeable wheel-shaped print mechanism on which characters are represented by embossed metal slugs at the ends of spokes. The daisywheel mechanism is used in printers manufactured by Diablo and Qume.

data bank—a comprehensive collection of data.

data base management system—a set of programs designed to organize, store, and retrieve machine-readable information from a computer-maintained data base or data bank.

dictation system—a configuration of equipment designed to record voice communications for subsequent transcription by a typist.

direct entry phototypesetter—a typesetter with a keyboard for the direct entry of text and commands. In effect, a direct entry phototypesetter is a text editing system that produces typeset rather than typewritten output.

diskette—a floppy disk.

display-oriented text editing system—in word processing, a text editing system which utilizes a CRT display with keyboard for the input of text. Information entered at the keyboard is displayed on the screen and subsequently can be printed or recorded on magnetic media.

display terminal—a terminal which displays information on a screen rather than printing it onto paper.

document assembly—in word processing, the ability to assemble documents out of sections of prerecorded text.

dot matrix—an array of points of ink, light, or similar image-forming elements that are used to form alphanumeric characters.

dry silver microfilm—a type of microfilm, used in source document cameras and COM recorders, which is sensitive to light but is developed by heat without the wet chemicals characteristic of conventional silver halide microfilm.

"dumb" terminal—a terminal without self-contained information processing capability, as opposed to an "intelligent" terminal which can be programmed for operation apart from any remote computer.

E

EBCDIC—Extended Binary Coded Decimal Interchange Code. An 8-bit computer code that is used to represent upper- and lowercase characters and special symbols.

electrofax—a variant of the electrostatic process in which images are created out of electrical charges on a specially treated paper and subsequently are developed by the application of a wet or dry toner. The electrofax process is used in office copiers and facsimile equipment.

electronic mail and message systems (EMMS)—a technique that includes equipment and communications technology that can permit entry, storage, and delivery of person-to-person messages.

F

facsimile—(1) an exact copy of a document. (2) the process or result by which fixed graphic images are scanned, transmitted electronically, and reproduced either locally or remotely.

fiber optics—very thin strands of glass that can be used to conduct information in the form of successive light pulses. Fiber optics are an increasingly important component in electronic transmission systems, computer printers, and copiers.

firmware—prewritten programs stored in Read-Only Memory (ROM) circuits. Firmware is used widely as a method of implementing commonly used software in word processing and small computer systems.

floppy disk—in word processing and small computer systems, a recording/storage medium consisting of a magnetic coated circular-shaped piece of polyester on which individual characters are recorded in concentric tracks. Floppy disks come in two sizes. The standard floppy measures 8 inches in diameter while the mini-floppy measures 5.25 inches. Character storage capacity varies from system to system, but rarely exceeds half a million characters per disk.

full duplex—a communication attribute in which a given device can transmit and receive information simultaneously.

G–H

global change—in word processing, the ability to correct or otherwise revise all occurrences of a given character, word, or phrase in one operation.

half duplex—a communication attribute in which a given device can transmit and receive information, but not simultaneously.

I

impact printer—an output unit that prints characters on paper by physical contact.

integrated information system—an information system that combines two or more technologies.

intelligent copier—a microprocessor-controlled device that can function as either an office copier or as a printer linked to a computer or word processing system.

intelligent terminal—a microprocessor-controlled device that is capable of functioning as a self-contained microcomputer or as a terminal communicating with a remote computer.

intelligent typewriter—an enhanced typewriter with the ability to perform certain difficult typing tasks automatically, such as centering of characters and decimal alignment.

K–L

keyboard send-receive terminal—a display or printing terminal that can both transmit data to, and receive data from, a computer or other information processing device. Transmission is initiated through a keyboard.

labor productivity—an economic measure of productivity calculated by dividing the Gross National Product by the number of persons or employee-hours in the workforce.

large-scale integration (LSI)—a manufacturing technique in which large numbers of highly miniaturized electronic circuits are combined or integrated on a single chip of silicon or other semiconductor material. LSI has permitted very significant reductions in the cost of microprocessor and memory circuits.

lines per inch—a measure of resolution in facsimile transmission.

list management—in word processing systems, the ability to process a prerecorded list of names or other items, selecting only those which meet certain specified criteria.

M

magnetic card—a magnetic-coated tabulating size card used as the recording/storage medium in some text-editing systems.

memory typewriter—an enhanced typewriter featuring the most basic text-editing capabilities and a nonremoveable magnetic recording medium with capacity for 50 to 100 pages of text.

microcomputer—a computer with a microprocessor as its central processing unit.

microfacsimile—the transmission and/or reception of microimages via facsimile communication.

microfilm camera/processor—the picture taking portion of a microfilming mechanism and the lens.

microform—a form, usually film, which contains microimages.

micrographics—techniques associated with the production, handling, and use of microforms.

microprocessor—a single chip integrated circuit device capable of performing the arithmetic and logical operations typically associated with a central processing unit in a computer system.

modem—a device that converts the electrical signal emitted by an information-processing machine to a form suitable for transmission over telephone lines or other communication facilities.

multifunction system—as applied to office automation, a system that can perform several types of information processing—for example, word processing, in combination with data processing and/or electronic message transmission.

N–O

nonimpact printer—a printing device in which the paper is not struck, but imaged by other means, e.g., ink jet, electrostatic.

one-line display—in word processing, a modified typewriter-based text editing system which includes a calculator-like display enabling the operator to view and revise approximately one line of typed input prior to printing.

online—(1) pertaining to equipment or devices under control of the central processing unit. (2) pertaining to a user's ability to interact with a computer. (3) pertaining to equipment or devices under control of a computer-output microfilmer.

online data base—an integrated accumulation of machine-readable data maintained on one or more direct access storage devices.

online terminal—a device that permits the interactive transmission of data to, or the reception of data from, a computer or other information-processing device via electronic digital pulses transmitted over connecting wires.

optical character recognition (OCR)—a technique by which printed or photographically recorded characters can be rapidly recognized by a combination of scanning techniques and electronic logic and converted to binary digital codes for storage, transmission, etc.

optical disk—a device that contains data (audio/video) recorded on spiral or circular tracks with a low-power laser.

P

page printers—computer output devices that use a combination of laser and xerographic technologies to print documents at a rate of two to five pages per second.

photocomposer—a phototypesetting device that features a make-up display terminal which both typesets text and composes pages.

phototypesetter—a nonimpact typesetting device in which character images are stored on an internal matrix and are exposed selectively onto a photosensitive paper or film master.

pixels—in facsimile transmission, the individual picture elements of the subject copy which are analyzed successively for their light reflectance values. An appropriate signal then is transmitted and directs the receiving device in the reconstruction of successive pixels.

playback—in word processing, the printing or display of previously recorded text.

plotter—a computer peripheral device designed to produce graphics output in paper form.

productivity—the relationship between the output of goods and services and the input of capital, labor, and natural resources.

Q–R

query language—in data base management systems, an alternative to conventional programming languages which allows users to formulate *ad hoc* information—retrieval requests without formal training in algorithmic thought.

random access memory (RAM)—semiconductor memory circuits used for the storage of data and programs in word processing systems, computer systems, and related information-processing devices.

read-only memory (ROM)—semiconductor memory circuits that contain prewritten programs or data. The content of ROM circuits is permanent, while the content of RAM circuits is volatile—that is, it is erased when the power supply to the circuits is interrupted.

real-time processing—a mode of computer processing in which information about a transaction or other event is processed at the time the event occurs, as opposed to batch processing in which there is an interval in time between the occurrence of an event and the processing of information about that event.

receive-only (RO) terminal—a display or printing terminal capable of receiving, but not transmitting, data. Physically, an RO terminal is distinguished by the absence of a keyboard.

recorder—in dictation systems, the machine that records spoken words as an analog signal on magnetic media.

reprographics—the art and science of reproducing documents.

run-length encoding—in facsimile systems, a technique used to compress a digital signal to increase the speed of transmission.

S

serial asynchronous transmission—a method of transmission in which the individual bits which encode a given character are transmitted in sequence and framed (preceded and followed) by additional bits which separate successively transmitted characters from one another.

shared logic system—in word processing, a text-editing system in which the central processor is designed to support multiple input and output devices.

shared resource system—in word processing, a text-editing system that combines stand-alone input stations capable of local operation and text storage with other components, such as printers or large capacity disk storage, which are shared by the various input stations. The purpose of shared resource systems is to maximize the utilization of infrequently used equipment components.

small business computer system—a microcomputer or small minicomputer configured with peripherals, and perhaps software, suitable for data processing applications.

small office microfilm (SOM)—a term used to denote a group of micrographics products designed specifically for small organizations or decentralized microfilming operations within larger organizations. SOM equipment is typically low in cost and designed for simplified use by nontechnical personnel in an office environment.

smart terminal—a microprocessor-controlled terminal that offers display and printing features not found on conventional "dumb" terminals. Unlike "intelligent" terminals, however, smart terminals are not programmable and have no integral information-processing capabilities.

software—a set of programs, procedures, and documents concerned with the operation of a data processing system.

source document microfilming—the conversion of documents, usually paper, to microimages.

stand-alone system—in word processing, a self-contained text-editing system that features a dedicated central processor designed to control one input and one output device.

synchronous transmission—a method of transmission in which sending and receiving devices are synchronized, thus eliminating the need for framing bits used in asynchronous transmission.

system software—programs that enable a computer to function and control its own operation, as opposed to application programs which accomplish some user-specified task. The most common example of system software is the group of programs described as the computer's operating system.

T

technological change—advances in technology that favorably influence the cost of producing goods and services.

teleconferencing—a technique that permits person-to-person messages, usually with an agenda and a group of communicators who may operate in real-time mode. Current systems can include video, audio, and computer technologies as well as communications.

teleprinter—a typewriter-like terminal designed to produce hardcopy output.

teletext—a form of videotext service in which information is transmitted over television broadcast channels, using otherwise blank portions of television frames. Information is assembled in a magazine format for transmission on a continuous sequential basis. The user consults a series of indexes to select the desired frame(s) for display.

teletypesetter (TTS) tape—a punched paper tape containing commands and characters required by a phototypesetter or photocomposer.

telex—a network of telegraph-grade lines and terminals designed for the interchange of domestic and international messages.

text-editing systems—in word processing, systems designed to capture keystrokes on magnetic media for subsequent printing, correction, revision, or other manipulation. The terms "text editor" and "word processor" are often used synonymously.

total productivity—an economic measure of productivity computed by dividing the Gross National Product by a combination of labor and capital, each weighted to reflect its relative contribution to the Gross National Product.

transcriber—in dictation systems, the machine that converts previously recorded signals to audible sounds for playback through a speaker or earphones.

turnkey system—an integrated configuration of preselected hardware and prewritten software designed to accomplish some information-processing task. The term is applied most often to dedicated computer systems employing minicomputers or microcomputers.

TWX—a network of telegraph-grade lines and terminals designed for the transmission of messages within North America.

typewriter-based text-editing system—a text-editing system that uses a modified typewriter as a combined input/output device.

typewriter quality printer—in word processing and computer systems, an output device capable of producing quality equivalent to that of an office typewriter.

U–V

updatable microfiche system—a microfilm that permits the addition or deletion of images.

value-added networks—alternative information transmission networks designed to facilitate the communication of information among terminals and computers. Most value-added networks offer long-distance communication rates that compare very favorably with those of the conventional telephone network. Examples of value-added networks include Telenet, Tymnet, Graphnet, and ITT Fax-Pak.

videoconferencing—the use of television monitors, cameras, and specially designed studios to conduct a conference between groups of persons in two or more geographically separate locations.

videotext—television-based information services that allow users to access publicly available data banks through modified television sets in homes or offices.

viewdata—a form of videotext service in which information is transmitted over telephone lines to specially modified television sets in homes or offices.

W

Winchester-type disk—in word processing and small computer systems, a hard surface disk often used as a high-capacity alternative to floppy disks.

word—in computer systems, the number of bits of information that a central processing unit can access or manipulate in a single operation.

word originator—in word processing, the creator of a textual communication.

word processing—a system for organizing people, procedures, and automated equipment to transfer information more efficiently from a spoken or recorded form to a written form.

word processor—an automated text-editing system.

X

xerographic process—the formation of a latent electrostatic image by action of light on a photoconducting insulating surface. The latent image may be made visual by a number of methods, such as applying charged pigmented powders or liquids that are attracted to the latent image. The particles either directly or by transfer may be applied and fixed to a suitable medium.

Notes

Chapter 1

1. This figure is derived from the Dartnell Institute in Chicago, which regularly calculates the cost of document production. In 1979, the cost of a business letter was $5.59.
2. The study was conducted by Booz, Allen and Hamilton and described by Harvey Poppell as reported in Marguerite Zientara, "Productivity Pegged to Office Automation," *Computerworld,* May 5, 1980, pp. 1, 4. See also: Harvey L. Poppell, "Planning for the Electronic Office of the Future," *Management Review* 67 (June 1978): 29–31.
3. Newspaper and magazine coverage of the critical problem of office space has been extensive. See, for example: Carter B. Horsley, "Office Space Gets Tighter," *New York Times,* Nov. 26, 1980, p. 35; James Carberry, "Real Estate Perks Up in Lower Manhattan," *Wall Street Journal,* Nov. 4, 1980, p. 46; Philip H. Dougherty, "Shrinkage Forecast in Agency Office Space," *New York Times,* June 4, 1980, p. 15; James Carberry, "More Office Tenants Becoming Landlords as Space Scarcity Pushes Rents to Records," *Wall Street Journal,* April 30, 1980, p. 16; Martin Shannon, "Office Space May Double in Cost," *Wall Street Journal,* Sept. 20, 1979, p. 1; "The Pressures on Office Space," *The Banker* 129 (Nov. 1979): 111–13ff; "The Cost of Having a Downtown Address," *Business Week,* June 16, 1980, pp. 64–65; R. L. Collins, "The Shrinking Square Foot," *Administrative Management* 38 (Oct. 1977): 26–27. On the close relationship between office location and the development of information systems, see P. W. Daniels, "Perspectives on Office Location Research," in *Spatial Patterns of Office Growth and Location,* ed. P. W. Daniels (London, 1979), pp. 1–28; Roger Pye, "Office Location and the Cost of Maintaining Contact," *Environment and Planning* 9 (1977): 149–68.
4. A selection of recent articles typical of such criticism includes: Paul A. Strassman, "The Office of the Future: Information Management for the New

215

Age," *Technology Review* 82 (Jan. 1980): 54–63; Don M. Avedon, "Goal of the 80s: Automating Offices to Spur Productivity," *Office* 9 (April 1980): 106, 108; J. J. Connell, "It's Time to Take a Hard Look at Office Productivity," *Management Review* 67 (June 1978): 29–31; F. A. Bell, "Assault on Office Operation Costs," *Administrative Management* 39 (Feb. 1978): 100; G. Grove, "Information Management in the Office of the Future," *Management Review* 68 (Dec. 1979): 47–50; J. B. LeBoutillier, "Office Automation Impacts Productivity," *Journal of Micrographics* 13 (Sept./Oct. 1980): 18; Wayne L. Rhodes, Jr., "Office of the Future: Fact or Fantasy?" *Infosystems* 27 (March 1980): 45, 48, 52–54; R. A. Gehmlich, "The Records Management Specialist in the Office of the Future," *Records Management Quarterly* 14 (Jan. 1980): 5–7; Robert M. Thompson, "The Paperless Agency of the Future," *Canadian Insurance* 85 (Jan. 1980): 32–40.

5. On the continued viability of the industrial engineering approach to office operations, see N. Dean Meyer, "The Role of Management Science in Office Automation," *Interfaces* 10 (Feb. 1980): 72–76; Nelson J. Cyr, "Improve Productivity through Computerized Memomotion," *Industrial Engineering* 11 (July 1979): 34–39; C. D. Sadlier, "The Office of the Future: New Challenges for Operational Research," *Omega* 8 (1980): 21–28.

6. C. Wright Mills, *White Collar* (New York, 1953), especially pp. 189–95, 204–212. Mills drew heavily on the pioneering work of William H. Leffingwell in his descriptions of office automation through the first half of the twentieth century. See, for example, Leffingwell, *Scientific Office Management* (Chicago, 1917); *The Office Through a Microscope* (Chicago, 1918); *Office Management: Principles and Practice* (Chicago, 1925); *Making the Office Pay* (Chicago, 1918); *The Fundamentals of Scientific Office Management* (New York, 1927); *The Office Appliance Manual* (Chicago, 1926).

7. Examples include: George R. Terry, *Office Automation* (Homewood, Ill., 1966); E. V. Grillo, *Control Techniques for Office Efficiency* (New York, 1963); Frank M. Knox, *Integrated Cost Control in the Office* (New York, 1958); E. P. Strong, *Increasing Office Productivity* (New York, 1962); George R. Terry, *Office Management and Control* (Homewood, Ill., 1975); Harry P. Cremach, *Work Study in the Office* (London, 1965); James Bayhylle, *Productivity Improvements in the Office* (New York, 1968).

8. For an excellent, brief overview, see Robert Propst, *The Office—A Facility Based On Change* (Elmhurst, Ill., 1978).

9. The importance of this distinction between "structured" and "unstructured" office operations was noted first by M. D. Zisman, "Office Automation—Revolution or Evolution," *Sloan Management Review* 19 (March 1979): 1–16.

10. For a recent study that acknowledges the influence of factory models on office automation, see Richard J. Matteis, "The New Back Office Focuses on Customer Service," *Harvard Business Review* 57 (March/April 1979): 146–59.

11. R. K. Martin, "Don't Overlook Clerical Productivity," *Industrial Engineering* 9 (Feb. 1977): 28–33; G. L. Hershey, "Clerical Productivity is Management's Responsibility," *Management World* 9 (Jan. 1980): 13–14; "Office Automation: a Challenge to Be Faced," *Modern Office and Data Management* 18 (Aug. 1979): 4–8.

12. The best known description of managerial work habits is Henry Mintzberg's

1975 article, "The Manager's Job: Folklore and Fact," which is conveniently reprinted in *Harvard Business Review—On Human Relations* (New York, 1979), pp. 104–24. The article is based on Mintzberg's longer study of five chief executives, *The Nature of Managerial Work* (New York, 1973). Similar studies, which confirm Mintzberg's view of the discontinuous and essentially reactive character of managerial/supervisory work, include: Rosemary Stewart, *Managers and Their Jobs* (London, 1967); Richard Tanner Pascale, "Communication and Decision Making Across Cultures: Japanese and American Comparisons," *Administrative Science Quarterly* 23 (1978): 91–110; Pascale, "Comparison of Selected Work Factors in Japan and the United States," *Human Relations* 33 (1980): 433–55; D. Pym, "Professional Mismanagement—the Skill Wastage in Employment," *Futures* 2 (1980): 142–50; P. D. Olson, "The Overburdened Manager and Decision-Making," *Business Horizons* 22 (May 1979): 28–32; J. L. J. Machin and L. S. Wilson, "Closing the Gap Between Planning and Control," *Long Range Planning* 12 (1979) 16–32; R. L. Daft and N. B. Macintosh, "New Approach to the Design and Use of Management Information," *California Management Review* 21 (1978): 82–92; J. R. Emshoff, "Experience-Generalized Decision-Making—the Next Generation of Managerial Models," *Interfaces* 8 (1978): 40–48; W. G. Ryan, "Management Practice and Research—Poles Apart," *Business Horizons* 20 (March 1977): 23–29; R. B. Higgins, "Reunite Management and Planning," *Long Range Planning* 9 (1976): 40–45; C. B. Hetrick, "The Manager's Job," *Harvard Business Review* 53 (Nov./Dec. 1975): 187. For a reflection of the popular awareness of this problem, see F. A. Weil, "Management's Drag on Productivity," *Business Week*, Dec. 3, 1979, p. 14.

13. A number of studies confirm the role of informal information sources in scientific and technical fields. See, for example: Richard S. Rosenbloom and Francis W. Wolek, *Technology and Information Transfer: a Survey of Practice in Industrial Organizations* (Boston, 1970); Saul Herner and Mary Herner, "Information Needs and Uses in Science and Technology," *Annual Review of Information Science and Technology* 2 (1967): 22–43; Thomas J. Allen, "Performance of Information Channels in the Transfer of Technology," *Industrial Management Review* 8 (1966): 87–98; Herbert Menzel, "Scientific Communication: Five Sociological Themes," *American Psychologist* 21 (1966): 999–1004.

14. See Mills, *White Collar*, pp. 189–95.

15. "The Potential for Telecommuting," *Business Week*, Jan. 26, 1981, pp. 94–98 describes current industrial experiments involving employees who work at home. Public awareness of the potential of home-based employment has been stimulated by such books as Alvin Toffler, *The Third Wave* (New York, 1980).

16. Since the eighteenth century, classical economists have taken a disparaging view of those workers who are not directly involved in the production of goods and services. Attempts to resolve the continuing uncertainty about white collar productivity are discussed in Walter Presnick, "Measuring Managerial Productivity," *Administrative Management* 61 (May 1980): 26–28; B. W. Cannon, "New Frontiers in Productivity Improvement: White Collar Workers," *Industrial Engineering* 11 (Dec. 1979): 34–37; L. Adkins, "Getting a Grip on White Collar Productivity," *Dun's Review* 114 (Dec. 1979): 120–22ff; E. Mandt, "Managing the Knowledge Workers of the Future," *Personal Journal* 57 (March

1978): 138–43, 162; D. B. Miller, "How to Improve the Performance and Productivity of the Knowledge Worker," *Organizational Dynamics* 5 (Winter 1977): 62; A. T. Kearney, Inc., "Managing in a Service Economy," *Journal of Systems Management* 26 (July 1975): 12–19; Ann M. DeVilliers, "Understanding White-Collar Productivity: How to Solve the Problem," *Journal of Micrographics* 13 (July/Aug. 1980): 43–46.

17. From 1974 through September 1979, productivity figures were reported in the *Chartbook on Prices, Wages, and Productivity*, published by the Bureau of Labor Statistics. Since the *Chartbook* ceased publication, the most convenient source for productivity information is the Current Statistics section of the *Monthly Labor Review*. Greater detail is provided in other Bureau of Labor Statistics publications, including *Productivity, Prices, and Price Indexes; CPI Detailed Report; Employment and Earnings;* and *Current Wage Developments*. The *Handbook of Basic Economic Statistics*, published monthly by the Bureau of Economic Statistics, is one of several available compilations of statistical data provided by the federal government.

18. Several excellent books and articles provide a clear and brief explanation of these productivity measures: Solomon Fabricant, *A Primer on Productivity* (New York, 1969); John W. Kendrick, *Understanding Productivity: an Introduction to the Dynamics of Productivity Change* (Baltimore, 1977); Edgar Weinberg, "Productivity: Defining the Game and the Players," *IEEE Spectrum* 15 (Oct. 1978): 34–39; William T. Morris, *Work and Your Future: Living Poorer, Working Harder* (Reston, Va., 1975); Ralph E. Winter, "Many Culprits Named in National Slowdown of Productivity Gains," *Wall Street Journal*, Oct. 21, 1980, pp. 1, 24.

19. For a comparison of productivity statistics, see Keith Daly and Wilbur Neff, "Productivity and Unit Labor Costs in Eleven Industrial Countries," *Monthly Labor Review* 101 (Nov. 1978): 11–18; Arthur Neef and Patricia Capdevielle, "International Comparisons of Productivity and Labor Costs," *Monthly Labor Review* 103 (Dec. 1980): 32–39; J. Fred Bucy, "Exploding a Few Myths about Productivity and Presenting a Formula for the Future," *IEEE Spectrum* 15 (Oct. 1978): 45. On the much publicized productivity gains in Japan, see T. Ozawa, "Japanese World of Work: an Interpretive Survey," *MSU Business Topics* 28 (Spring 1980): 45–55; W. L. Givens and W. V. Rapp, "What it Takes to Meet the Japanese Challenge," *Fortune*, June 18, 1979, pp. 104–20; J. L. Riggs and K. K. Sed, "Personnel Factor in Japanese Productivity," *Industrial Engineering* 11 (April 1979): 32–35; M. Nishimizu and C. R. Hulten, "The Sources of Japanese Economic Growth, 1955–71," *Review of Economics and Statistics* 60 (Aug. 1978): 351–61; Ezra Vogel, *Japanese as Number One* (Cambridge, Mass., 1979). On the problems of comparing the productivity of different countries, see Joel Dean, "International Productivity Comparisons for Managerial Decisions," in *Labor Productivity*, eds. John T. Dunlop and V. P. Diatchenko (New York, 1964), pp. 177–85.

20. Seymour Melman, "The Rise of Administrative Overhead in the Manufacturing Industries of the United States, 1899–1947," *Oxford Economic Papers* 3 (1951): 62–93 provides a history of the growth of the nonproduction workers in the secondary sector. See also: John E. Ullman, "Administrative Overhead: Growth and Efficiency," in *Problems in the Growth and Efficiency of Administrative and Service Functions*, ed. Ullman (Hempstead, N.Y., 1978), pp. 1–5.

21. Leopold Froehlich, "Robots to the Rescue," *Datamation* 27 (Jan. 1981): 84–98; "Robots Join the Labor Force," *Business Week*, June 9, 1980, pp. 62–76; Fred Reed, "The Robots are Coming, the Robots are Coming," *Next* 1 (May/June 1980): 30–39.
22. The definitive delineation of the scope of the services sector is from Victor R. Fuchs, *The Service Economy* (New York, 1968); see also: *Production and Productivity in the Service Industries*, ed. Fuchs (New York, 1965).
23. The phrase was popularized by Daniel Bell, *The Coming of Post-Industrial Society* (New York, 1973).
24. D. Quinn Mills, "Human Resources in the 1980s," *Harvard Business Review* 57 (July/Aug. 1979): 154–63; Howard N. Fullerton, Jr., "The 1995 Labor Force: a First Look," *Monthly Labor Review* 103 (Dec. 1980): 11–21.
25. Representative discussions of the nontechnological approaches to office productivity include: P. C. Hughes, "Lighting, Productivity, and the Work Environment," *Lighting Design and Application* 8 (Dec. 1978): 32–40; J. F. Barnaby, "Lighting for Productivity Gains," *Lighting Design and Application* 10 (Feb. 1980): 20–28; H. H. Young and G. L. Berry, "Impact of Environment on the Productivity Attitudes of Intellectually Challenged Office Workers," *Human Factors* 21 (Aug. 1979): 399–407; J. S. Prince, "Environments that Work for People," *Administrative Management* 41 (Oct. 1980): 36, 60ff; M. Magnus, "Office Automation, Personnel, and the New Technology," *Personnel Journal* 59 (Oct. 1980): 815–23; J. B. Nelson, "Improving Productivity with Performance Measures and a Group Incentive Plan: a Case History," *Proceedings of the AIEE Annual Conference* (Norcross, Ga., 1977), pp. 411–13; J. T. Fucigna, "The Ergonomics of the Office," *Ergonomics* 10 (1967): 589–604.
26. The classic explanation is Edwin Mansfield, *The Economics of Technological Change* (New York, 1968). It is available in a shorter version entitled *Technological Change* (New York, 1971). See also: Edward Denison, *The Sources of Economic Growth in the United States and Alternatives Before Us* (New York, 1962); W. E. G. Salter, *Productivity and Technological Change* (Cambridge, 1965); R. Solow, "Investment and Technological Progress," in *Mathematical Models in the Social Sciences*, eds. K. J. Arrow, S. Karlin, and P. Suppes (Stanford, 1960); Murray Brown, *On the Theory and Measurement of Technological Change* (Cambridge, 1968).
27. For a clear summary of the problem of declining capital investment, see Campbell R. McConnell, "Why is U.S. Productivity Slowing Down?" *Harvard Business Review* 57 (March/April 1979): 36–60.
28. Pessimistic proponents of this viewpoint include Daniella H. Meadows and D. H. Meadows, *The Limits to Growth* (New York, 1972); Edward F. Renshaw, *The End of Progress* (North Scituate, Mass., 1976).
29. Marc Porat, *The Information Economy* (Washington, D.C., 1977); Porat, "Global Implications of the Information Society," *Journal of Communication* 28 (1978): 70–80.

Chapter 2

1. For a comparatively new field, there is a surprisingly large and rapidly growing literature on word processing. Recent books include: Walter A. Kleinschrod,

L. B. Kruk, and H. J. Turner, *Word Processing: Operations, Applications, and Administration* (Indianapolis, 1980); Paula B. Cecil, *Management of Word Processing Operations* (Reading, Mass., 1980); Rita Kutie and Virginia Huffman, *The Word Processing Book* (New York, 1980); and Shirley Waterhouse, *Word Processing Fundamentals* (New York, 1978). The phenomenal growth of word processing is reflected in the development of several specialized magazines and newsletters, including *Word Processing and Information Systems* (formerly *Word Processing Systems* and, before that, *Word Processing World*), the *Information and Word Processing Report,* and the *International Word Processing Report,* all of which are published by Geyer-McAllister. Word processing survey articles and case studies appear regularly in business and trade publications. While a complete bibliography is beyond the scope of this book, some representative recent examples include: W. A. Walshe, "Text Editing Machines for Differing Applications," *Administrative Management* 41 (June 1980): 42–46, 70, 75; Thomas Holmes, "The Latest Word on Word Processors," *Output* 1 (Sept. 1980): 38–40, 53, 55; Amy Wohl, "A Review of Office Automation," *Datamation* 26 (Feb. 1980): 116–19; J. E. Smith, "Choosing a New Generation Word Processing System," *Office* 92 (Nov. 1980): 133–34, 138; Victor J. Krasan, "Enter the Electronic Editor," *Management Focus* 27 (Nov./Dec. 1980): 28; Randy J. Goldfield, "Word Processing Boom," *Dun's Review* 122 (1978): 14; F. Greenwood, "Word Processing Primer," *Journal of Systems Management* 29 (1978): 36–38; Denis Ryan, "Technological Change in the Tertiary Sector—Three Examples," *Work and People* 5 (1980): 13–18; R. W. Ketron, "Four Roads to Office Automation," *Datamation* 26 (Nov. 1980): 138–40. In addition, many articles examine the impact of word processing on particular occupations—for example: T. J. Dixon, "Word Processing for the Appraiser," *Real Estate Appraiser and Analyst* 46 (Sept./Oct. 1980): 7–9; "The Office: Equipment is Changing the Face of Government," *Government Executive* 12 (Oct. 1980): 14, 16; "Stock Analysis Speeded Up with Word Processing System," *Infosystems* 27 (Oct. 1980): 116, 118; Wallace Richman, "New Technology in Word Processing," *New York Law Journal,* Nov. 3, 1980, p. 4; P. M. McClellan, "Word Processing Applications in the Seattle Police Department," *Police Chief* no. 4 (1980): 74; C. A. Ashland, "Employing Blind Persons in Word Processing Centers," *Journal of Visual Impairment and Blindness* 73 (1979): 293; D. E. Stepien et al., "Word Processing in the 1980s," *College and University* 55 (1980): 402; L. B. Kruk, "Word Processing and its Implications for Business Communications Courses," *Journal of Business Communication* 15 (1978): 9–18; K. Letcher and D. A. Pierino, "Word Processing Equipment for Hospital Pharmacy Applications," *American Journal of Hospital Pharmacy* 36 (1979): 1529–33; G. E. Dower et al., "Standardization of Electrocardiographic Interpretive Statements—Menu for Word Processing," *Canadian Medical Journal* 120 (1979): 808–12; C. T. Ashworth et al., "Computerized Word Processing and Data Systems for Histology in a Private Medical Laboratory," *American Journal of Clinical Pathology* 71 (1979): 257–62; L. W. Moulton, "Word Processing Equipment for Information Centers," *Special Libraries* 71 (1980): 492–96; "Paperwork Problems: A Small Life Agency Finds Word Processing Helpful," *Canadian Insurance* 85 (Sept. 1980): 30–31.

2. On the opportunities for productivity improvement inherent in dictation systems, see J. D. Gould and S. J. Boies, "How Authors Think About Their Writing,

Dictating, and Speaking," *Human Factors* 20 (1978): 495–505; also Gould and Boies, "Writing, Dictating, and Speaking Letters," *Science* 201 (1978): 1145–47, reprinted in *IEEE Transactions on Professional Communications* 22 (1979): 16–18. For a good economic analysis, see Vincent J. Laraia, "Justifying Dictation: One User's View," *Modern Office Procedures* 25 (Dec. 1980): 86–93.
3. George Leslie, "Variable Speech Technology for Dictation Equipment," *Word Processing Report,* June 1, 1978, unpaginated; "How Variable Control Eases Playback," *Word Processing World* 5 (Sept. 1978).
4. As this book is written, this capability is available on dictation equipment manufactured by Sony.
5. For a review of current developments in voice entry of data and text, see: George Brandt, "Speech Recognition—Machines that Talk and Listen," *Telephone Engineering and Management,* April 1, 1980, pp. 74, 77, 80–82; E. J. Simmons, Jr., "Voice—A Solution to the Data Entry Bottleneck," *Advances in Instrumentation* 34 (1979): 195–201; F. Jelinek, "Continuous Speech Recognition for Text Applications," in *Electronic Text Communication,* ed. W. Kaiser (Berlin, 1978), 262–76; Jake Kirschner, "Glaser Says Voice Recognition Still in Infancy," *Computerworld,* March 24, 1980, p. 15; "Tips Given on Speech Recognition Needs," *Computerworld,* Feb. 11, 1980, pp. 57, 62; M. Pelters, "Time to Have Words with your Computer," *International Management* 34 (Nov. 1979): 23–25; E. G. Nassembene, "Speech Input Technique for Data Entry to Word Processing Equipment," *IBM Technical Disclosure Bulletin* 22 (1979): 2891–92; A. Komiyama et al., "A Monosyllabic Voice Typewriter with Learning Process," *Bulletin of Research of the Institute of Applied Electronics* 29 (March 1977): 21–39; H. A. Elder, "On the Feasibility of Voice Input to an Online Computer Processing System," *Communications of the Association for Computing Machinery* 13 (1970): 339–46.
6. For a review of printing technology, see: George Rea, "Making Printers Work Effectively," *Small Systems World* 8 (Sept. 1980): 20–23, 35; Irving L. Wieselman, "Non-Impact Printing in the 1980s," *Mini-Micro Systems* 13 (Jan. 1980): 93–100; Jeffrey S. Prince, "The Pros and Cons of Ink Jet Printers," *Administrative Management* 41 (July 1980): 55–57; Irving L. Wieselman, "State of the Art Report: Line Printers," *Microcomputer News,* Oct. 11, 1979, pp. 1, 14, 23; David Whieldon, "The New Freedom in Computer Output—Part II: Printers and Plotters," *Computer Decisions* 11 (Aug. 1979): 32–34, 38–47; Stephen A. Caswell, "The Changing World of Nonimpact Printing," *Datamation* 24 (Nov. 1978): 128–32; W. L. Buehner et al., "Application of Ink Jet Technology to a Word Processing Output Printer," *IBM Journal of Research and Development* 21 (Jan. 1977): 2–9; Christopher Baker, "Ink Jet Printing—Its Principles and its Potential for Newspapers," *Production Journal* 72 (July 1976): 41, 43.
7. There have been several studies of the desirable attributes of word processing displays: F. Jones, "Using Word Processing Screens," *Display Technology and Applications* 1 (Oct. 1979): 149–53; K. N. Bobo, "System for Obtaining Display Attributes," *IBM Technical Disclosure Bulletin* 22 (March 1980): 4639–40; J. W. Rice and J. A. Volten, "Word Processing Terminal with Dual Display," *IBM Technical Disclosure Bulletin* 22 (March 1980): 4639–40; P. R. Herrold and H. C. Tanner, "Alphanumeric CRT Display having a Plurality of Display Positions," *IBM Technical Disclosure Bulletin* 22 (Sept. 1979): 1338–39; D. A. Bishop, "New Approach to Display of Graphics in Text Processing Systems," *IBM Technical*

Disclosure Bulletin 22 (Sept. 1979): 1331–33; G. Briscoe, "Human Considerations in Word Processing," in *Word Processing: Selection, Implementation, and Uses* (London, 1979), 261–72; H. O. Holt and F. L. Stevenson, "Human Performance Considerations in Complex Systems," *Journal of Systems Management* 29 (Oct. 1978): 14–20. On the relationship between the use of CRT displays and worker fatigue, see: S. Rosenthal, "Ocular Problems of VDU's: Fact and Fiction," in *Word Processing: Selection, Implementation, and Uses,* 131–44; O. Ostberg, "The Health Debate," *Reprographics Quarterly* 12 (Summer 1979): 80–83; William B. Knowles and Joseph W. Wulfeck, "Visual Performance with High-Contrast Cathode-Ray Tubes at High Levels of Ambient Illumination," *Human Factors* 14 (1972): 521–32; John D. Gould, "Visual Factors in the Design of Computer-Controlled CRT Displays," *Human Factors* 10 (1968): 359–76; O. Ostberg, "Fatigue in Clerical Work with CRT Display Terminals," *Goteborg Psychological Reports* 4 (1974): 1–14; G. V. Hultgren and B. Knave, "Discomfort Glare and Disturbances from Light Reflections in an Office Landscape with CRT Display Terminals," *Applied Ergonomics* 5 (March 1974): 2–8.

8. For descriptions of available OCR readers, see W. A. Walshe, "Optical Character Recognition: High Speed Input Increases Reliability," *Word Processing World* 6 (May 1979): 18–24; "Inside OCR: Saving Keystrokes and Money," *Modern Office Procedures* 25 (July 1980): 48–60; Steven Stebbins, "Optical Character Recognition: Rediscovering an Old Technology," *Infosystems* 27 (June 1980): 76–78; Herbert F. Schantz, "OCR Cuts Labor Costs, Boosts Office Efficiency," *Computerworld,* April 28, 1980, pp. 15–17; B. Bambrough, "Optical Character Readers in the Automated Office," *Computers and People* 29 (1980):19; G. L. Goodrich et al., "Kurzweil Reading Machine—Partial Evaluation of its Optical Character Recognition Error Rate," *Journal of Visual Impairment* 10 (1979): 389–99.

9. For a description of typical capabilities, see W. A. Walshe, "Math-Sort Packages—New Dimensions in Word Processing," *American Bar Association Journal* 65 (Jan. 1979): 72; E. A. O'Neill, "Word Processing Simplifies the Preparation of Financial/Accounting Reports and Statistical Tabulations," *National Public Accountant* 25 (July 1980): 21–23; "Using a Word Processor for Advanced Accounting," *Modern Office and Data Management* 19 (Aug. 1980): 14–16; "How Well Does DP Write and WP Compute," *Modern Office Procedures* 25 (May 1980): 71–84; Sally M. Saffer and Myles Walsh, "Word Processing vs. Data Processing," *Administrative Management* 41 (June 1980): 36–41, 66–67, 74; Jack L. Balderston, "WP and DP: a Compatible Couple?" *Management World* 9 (May 1980): 31–32; Paul H. Cheney, "You can Merge DP and WP," *Computer Decisions* 12 (Oct. 1980): 110, 112; Bruce Hoard, "Panelists Point Out Obstacles in Merger of DP, WP Technologies," *Computerworld,* May 26, 1980, p. 38; Werner L. Frank, "What are We Integrating with WP and DP?" *Computerworld,* Nov. 10, 1980, pp. 49–50.

Chapter 3

1. There have been a number of recently published books dealing with data communications: John E. McNamara, *Technical Aspects of Data Communications*

(Bedford, Mass., 1977); Daniel R. McGlynn, *Distributed Processing and Data Communications* (New York, 1978); D. W. Davies et al., *Computer Networks and their Protocols* (New York, 1979); Dimitris N. Chorafas, *Data Communication for Distributed Information Systems* (New York, 1980).

2. Useful books on data base concepts and systems include: C. J. Date, *An Introduction to Database Systems*, 2nd ed. (Reading, Mass., 1977); David Kroenke, *Database: a Professional's Primer* (Chicago, 1978); William H. Inmon, *Effective Data Base Design* (Englewood Cliffs, N.J., 1981); James Martin, *Computer Data Base Organization* (Englewood Cliffs, N.J., 1977); Martin, *Principles of Data Base Management* (Englewood Cliffs, N.J., 1976); Ronald G. Ross, *Data Base Systems: Design, Implementation, and Management* (New York, 1978); Harry Katzan, *Computer Data Management and Data Base Technology* (New York, 1975). There are a number of articles that deal with the management advantages of data base technology: C. J. Lewis, "Understanding Database and Data Base," *Journal of Systems Management* 28 (Sept. 1977): 36; G. C. Everest, "Database Management Systems Tutorial," *Proceedings of the American Institute for Decision Sciences* (Nov. 1975): 225; R. L. French, "Making Decisions Faster with Data Base Management Systems," *Business Horizons* 23 (Oct. 1980): 33–36; N. Ahituv and M. Hadass, "Identifying the Need for a DBMS," *Journal of Systems Management* 31 (Aug. 1980): 30–33; H. B. Josephine, "The Information Edge—How You Can Gain the Advantage," *Journal of Applied Management* (May/June 1980): 26–27; A. Vazsonyi, "Through the Looking Glass to Data Base Management," *Interfaces* 9 (Nov. 1979): 99–104.

3. Raymond McLeod, *Management Information Systems* (Chicago, 1979); Jerome Kanter, *Management-Oriented Management Information Systems* (Englewood Cliffs, N.J., 1977); Gordon B. Davis, *Management Information Systems: Conceptual Foundations, Structure, and Development* (New York, 1974); William A. Bocchino, *Management Information Systems, Tools and Techniques* (Englewood Cliffs, N.J., 1972); William C. House, ed., *Interactive Decision-Oriented Data Base Systems* (New York, 1977); Henry C. Lucas, *Computer-Based Information Systems in Organizations* (Chicago, 1973); Lucas, *Why Information Systems Fail* (New York, 1975); M. E. Walsh, "MIS—Where are We, How Did We Get Here, and Where are We Going?" *Journal of Systems Management* 29 (Nov. 1978): 6–21; James C. Emery, "Concepts of Management Information Systems," *Automatic Data Processing Handbook* (New York, 1977): III, 17, 28.

4. The most comprehensive survey is Richard Matick, *Computer Storage Systems and Technology* (New York, 1977).

5. For a relatively nontechnical explanation, see George C. Kenney et al., "An Optical Disk Replaces Twenty-Five Magnetic Tapes," *IEEE Spectrum* 16 (Feb. 1979): 33–38; T. Scannell, "Optical Memory Disk Challenging Tape and Disk," *Computerworld*, June 23, 1980, p. 65; "Digital Disks Predicted Next Mass Storage Technology," *Computerworld*, March 24, 1980, pp. 43, 50; Esther Surden, "Optical Mass Storage Expected to Rival Disk Storage," *Computerworld*, March 14, 1977, p. 43. More technical analyses include: Gerard O. Walter, "Will Optical Disk Memory Supplant Microfilm?" *Journal of Micrographics* 13 (July/Aug. 1980): 29–34; Bulthuis Kees et al., "Ten Billion Bits on Disk," *IEEE Spectrum* 16 (Aug. 1979): 26–33; William C. Donelson, "Spatial Management of Information," *Computer Graphics* 12 (1978): 203–209; A. H. Firester, "Optical Principles of Information Retrieval from Optical and Non-

Optical Recordings," in *SPIE Seminar Proceedings: Optical Information Storage* (Bellingham, Wash., 1979), 115–21; D. Vincent and J. W. Y. Lit, "Optical Disk in Thin Films," *Canadian Journal of Physics* 57 (1979): 1309–18; G. R. Knight, "Holographic Memories," *Optical Engineering* 14 (1975): 453–59.

6. The problem is summarized in "Missing Computer Software," *Business Week*, September 1, 1980, pp. 46–53. See also: Martin A. Goetz, "The Software Package Generator," *Computerworld*, Oct. 6, 1980, pp. 27–32.

7. With regard to the potential business significance of data banks, see D. R. Guillet, "On-Line Data Banks," *Administrative Management* 41 (July 1980): 44–47; Samuel A. Wolpert, "Management Role for Researcher to Grow as Use of Information Retrieval Systems Expands," *Marketing News*, Sept. 9, 1977, pp. 1, 14; Jay Gould, "Panning for Gold from Torrents of Data," *Sales and Marketing Management*, June 13, 1977, pp. 104–106; "Data Banks Blast Off," *Industry Week*, Dec. 6, 1976, pp. 42–43; W. Kiechel, "Everything You Always Wanted to Know May Soon be Online," *Fortune*, May 5, 1980, pp. 226–28ff.

8. For a more detailed discussion of these news-oriented information sources, see R. Slade and A. M. Kelly, "Sources of Popular Literature Online—*New York Times* Information Bank and the *Magazine* Index," *Database* 2 (1979): 70–83; Donna R. Dolan, "Subject Searching of the *New York Times* Information Bank," *Online* 2 (1978): 26–30; John Rothman, "*New York Times* Information Bank," *Special Libraries* 63 (1972): 111; Rhoda Garoogian, "Library Use of *New York Times* Information Bank—Preliminary Survey," *RQ* 16 (1976): 59–64; J. C. Moulton, "Dow Jones News-Retrieval," *Database* 2 (1979): 54–64.

9. For an historical review of online bibliographic data bases, see F. W. Lancaster and E. G. Fayen, *Information Retrieval On-Line* (Los Angeles, 1973), especially pp. 63–125; Carlos Cuadra, "Commercially Funded On-Line Retrieval Services: Past, Present, and Future," *Aslib Proceedings* 30 (1978): 2–15; Beatrice Marron and Dennis Fife, "On-Line Systems—Techniques and Services," in *Annual Review of Information Science and Technology*, ed. M. E. Williams (Washington, 1976), 163–201; J. L. Hall, comp., *On-Line Information Retrieval, 1965–1976: A Bibliography with a Guide to On-Line Data Bases and Systems* (London, 1977).

10. The literature on videotext services is growing rapidly. See, for example: Efrem Sigel, ed., *Videotext: the Coming Revolution in Home/Office Retrieval* (White Plains, N.Y., 1980); E. M. Ferrarini, "The Viewdata Experience," *Administrative Management* 41 (Oct. 1980): 20–25; "The Ultimate TV Set: It Shops, Buys, Teaches," *US News and World Report*, April 14, 1980, pp. 78–79; L. J. Anthony, "Viewdata," *Online* 3 (Oct. 1979): 73–75; Sandra K. Paul, "Viewdata: a New Way of Distributing Books and Information," *Publishers Weekly*, Jan. 8, 1979, pp. 31–32; Lois F. Lunin, "Data Bases + Television + Telephone = Viewdata," *Bulletin of the American Society for Information Science* 5 (Oct. 1978): 22–24; "Viewdata: A Review and Bibliography," *Online Review* 2 (1978): 217–24.

11. H. Takeuchi and Allan H. Schmidt, "New Promise of Computer Graphics," *Harvard Business Review* 58 (Jan./Feb. 1980): 122–31 provides a readable, management-oriented introduction with interesting illustrations.

12. As explained by K. E. Knight and R. P. Cerveny in *The Encyclopedia of Computer Science* (New York, 1976), p. 599, Grosch's Law states that "computing power increases as the square of the cost of the computer or $P = KC^2$ where P = computing power, K = a constant, and C = system cost (either lease price

or purchase price), so that, for example, for twice the money one obtains four times the computing power." For a critique, see E. G. Cale, L. L. Gremillion, and J. L. McKenney, "Price/Performance Patterns of U.S. Computer Systems," *Communications of the Association for Computing Machinery* 22 (April 1979): 225–33; Frank V. Wagner, "Is Decentralization Inevitable?" *Datamation* 22 (Nov. 1976): 86–97; Paul La Voie, "Intelligent Terminal—Defense Against Murphy's Law," *Computerworld*, March 28, 1977, pp. 18–19.

13. Charles Wojslaw, *Integrated Circuits: Theory and Applications* (Reston, Va., 1978) provides a brief overview of the technology. For an historical analysis, see Ernest Brown and Stuart MacDonald, *Revolution in Miniature* (Cambridge, 1977). On future directions in integrated circuit design and the emergence of Very Large Scale Integration (VLSI), see Daniel Queyssac, "Projecting VLSI's Impact on Microprocessors," *IEEE Spectrum* 16 (May 1979): 38–41; Jean-Michel Gabet, "VLSI: the Impact Grows," *Datamation* 25 (June 1979): 108–13; Arthur L. Robinson, "Are VLSI Microcircuits too Hard to Design?," *Science* 209 (July 1980): 258–63; "The Chip Makers Glamorous New Generation," *Business Week*, Oct. 6, 1980, pp. 117, 122ff.

14. Useful surveys include: Jack R. Buchanan and Richard G. Linowes, "Understanding Distributed Data Processing," *Harvard Business Review* 58 (July/Aug. 1980): 143–53; Buchanan and Linowes, "Making Distributed Data Processing Work," *Harvard Business Review* 58 (Sept./Oct. 1980): 143–61; G. Burnett and R. Nolan, "At Last, Major Roles for Minicomputers," *Harvard Business Review* 53 (May/June 1975): 148–56; A. L. Kelsch, "Dispersed and Distributed Data Processing," *Journal of Systems Management* 29 (March 1978): 32–37; A. Hoffberg, "Coming to Grips with Distributed Processing," *Administrative Management* 39 (May 1978): 52–68; Hamish Donaldson, *Designing a Distributed Processing System* (New York, 1979). Donald P. Kenney, *Minicomputers: Low Cost Computer Power for Management* (New York, 1978), a revised edition of the first book about minicomputers, is a good, nontechnical introduction. More technical presentations include: Richard H. Eckhouse, Jr. and L. Robert Morris, *Minicomputer Systems: Organization, Programming, and Applications (PDP-11)*, 2nd ed. (Englewood Cliffs, N.J., 1979); Donald Eadie, *Minicomputers: Theory and Operation* (Reston, Va., 1979); Martin Healey, *Minicomputers and Microprocessors* (New York, 1976); George D. Kraft, *Mini/Microcomputer Hardware Design* (Englewood Cliffs, N.J., 1979); John C. Cluley, *Programming for Minicomputers* (New York, 1978).

15. Microcomputers are discussed in a number of recent books, including: John D. Lenk, *Handbook of Microprocessors, Microcomputers, and Minicomputers* (Englewood Cliffs, N.J., 1979), a relatively technical, but very clear introduction; Gerald A. Silver, *Small Computer Systems for Business* (New York, 1978); Claire Summer and Walter A. Levy, *The Affordable Computer: the Microcomputer for Business and Industry* (New York, 1979); Jefferson C. Boyce, *Microprocessor and Microcomputer Basics* (Englewood Cliffs, N.J., 1979); Sol Liebes, *Small Computer Systems Handbook* (Rochelle Park, N.J., 1978); G. Rao, *Microprocessors and Microcomputer Systems* (New York, 1978); R. Tocci, *Microprocessors and Microcomputers* (Englewood Cliffs, N.J., 1979); Carol A. Ogden, *Microcomputer Design* (Englewood Cliffs, N.J., 1978); J. W. Willis, *Peanut Butter and Jelly Guide to Computers* (Portland, Ore., 1978), as the name implies, a nontechnical introduction.

16. Carol A. Ogden, *Software Design for Microcomputers* (Englewood Cliffs, N.J.,

1978) is an introduction to the programming of small computer systems. See also William T. Barden, *How to Program Microcomputers* (Indianapolis, 1977); Daniel D. McCracken, *A Guide to PL/M Programming for Microcomputer Applications* (Reading, Mass., 1978); Thomas McIntire, *Software Interpreters for Microcomputers* (New York, 1978); Peter Eichhorst, "System Software for Business Microcomputers," *Small Systems World* 7 (Dec. 1979): 22–23; M. E. Richerson, "Pascal Programming Language: Easy to Write and Trouble-shoot," *Machine Design* 52 (Aug. 1980): 112–18; M. J. McGowan, "High Level Microcomputer Languages Slash Software Development Costs," *Control Engineering* 27 (April 1980): 53–58; M. Schindler, "Pick a Computer Language that Fits the Job," *Electronic Design*, July 19, 1980, pp. 62–70, 72–78.

Chapter 4

1. For a review of the technology and typical applications, see: Daniel M. Costigan, *Micrographic Systems*, 2nd ed. (Silver Spring, Md., 1980); William Saffady, *Micrographics* (Littleton, Colo., 1978); Charles Smith, *Micrographics Handbook* (Dedham, Mass., 1978); and two older, but still useful works: G. W. W. Stevens, *Microphotography: Photography and Photofabrication at Extreme Resolution*, 2nd ed. (New York, 1968); and Carl Nelson, *Microfilm Technology: Engineering and Related Fields* (New York, 1965).
2. The close relationship between micrographics and office automation is reflected in a number of recent articles, including: E. T. Keating, "Productivity and Information Management," *Journal of Micrographics* 13 (July/Aug. 1980): 16–24, reprinted in *IMC Journal* 16 (3rd Quarter, 1980): 8–10; Charles I. Norris, "Integrated Office Systems: the Paperless Office," *Journal of Micrographics* 13 (May/June 1980): 23–26; Lincoln Hallen, "Integrating Micrographics into Future Office Systems," *Journal of Micrographics* 13 (March/April 1980): 71–77; Don M. Avedon, "Micrographics in the 1980s—a Technological Assessment," *IMC Journal* 16 (2nd Quarter, 1980): 19–22; "Will Micrographics Be the Productivity Savior?" *Modern Office Procedures* 25 (April 1980): 47–62; Joel Slutzky, "The Role of the Micrographics Industry in the Office of the Future," *Journal of Micrographics* 11 (Sept. 1977): 71–74.
3. On the technology of thermally processed microfilms, see: J. W. Shepard, "Dry Silver Films," *Proceedings of the National Microfilm Association* 22 (1973): 237–40; J. A. Norcross and P. I. Sampath, "Non Silver Photographic Materials and Lasers in the Micrographic Industry," in *Micrographics Science 1973: Winter Symposium*, ed. D. Chenevert (Washington, 1973); Kenneth R. Kurttila, "Dry Silver Film Stability," *Journal of Micrographics* 10 (1977): 113–19.
4. The available systems are described in: Col. (Ret.) Leonard S. Lee, "Everything You Wanted to Know about Updatables but Had No One to Ask," *Journal of Micrographics* 12 (Jan./Feb. 1979): 187–97; Thomas E. Burney, "Updatable Microforms," *Journal of Micrographics* 12 (July/Aug. 1979): 351–54.
5. The term was introduced in Stanley Nathanson and Richard Van Auken, "The Emergence of Small Office Microfilm (SOM) Systems," *Information and Records Management* 4 (Dec. 1970): 36–40. See also: John Van Auken and Richard Van

Auken, "Small Office Microfilm (SOM) Products: A Status Report," *Journal of Micrographics* 5 (1971): 5–11; William Saffady, "Small Office Microfilm (SOM) Products: A Survey for Libraries," *Microform Review* 5 (1976): 265–71; Alan W. Wilbur, "Microfiche Systems for the Small User," *Journal of Micrographics* 5 (1972): 287–89.

6. The fundamentals of COM are described in several sources, including: Robert F. Gildenburg, *Computer-Output-Microfilm Systems* (Los Angeles, 1974); William Saffady, *Computer Output Microfilm: Its Library Applications* (Chicago, 1978); S. H. Boyd, "Technology of Computer Output Microfilm: Past, Present, and Future," *TAPPI* 56 (1973): 107–10.

7. Mel Mandell, "The New Freedom in Computer-Output—Part I: Page Printers and Intelligent Copiers," *Computer Decisions* 11 (Aug. 1979): 22–24, 28; David Goodstein, "Output Alternatives," *Datamation* 26 (Feb. 1980): 122–30; F. W. Miller, "New Roles for Page Printers," *Infosystems* 25 (June 1978): 90–92; Ronald R. Weeks, "Side-Stepping Nonimpact Printer Problems," *Computer Decisions* 11 (March 1979): 52–57; Judith S. Hurwitz, "Price Not Only User Consideration When Choosing Nonimpact Printer," *Computerworld*, March 31, 1980, pp. 55, 59; "Squeezing Makers of Paper Forms," *Business Week*, June 9, 1980, p. 84; B. J. Shepherd, "Experimental Page Make-Up of Text with Graphics on a Raster Printer," *IBM Systems Journal* 19 (1980): 345–55.

8. This distinction, popular during the mid-1970s, is made in Tom L. Harrison, "CRT vs. COM—Real Time vs. Real Time Enough," *Journal of Micrographics* 9 (1975): 3–11; Frank Malabarba, "Microfilm Information Systems (MIS): A Data Base Alternative," *Journal of Micrographics* 9 (1975): 3–11

9. For a description of CIM systems, see Alfred L. Fenaughty, "Demand Printing: A Revolution in Publishing," *Journal of Micrographics* 8 (1975): 201–6; Arnold K. Griffiths, "From Gutenberg to Grafix I—New Directions in OCR," *Journal of Micrographics* 9 (1975): 81–89; Roger Holland, "CIM—the Present and the Future," *Microdoc* 15 (1976): 52—55; "OCR Solves Huge British Data Entry Problem," *Computerworld*, April 24, 1978, pp. 21, 25; McRae Anderson, R. J. Varson, and E. E. Wilkins, "Micrographics and Computer Technology Joined to Process 1980 Census Data," *Journal of Micrographics* 13 (July/Aug. 1980): 37–40.

10. On the history and technology of CAR systems, see: Vannevar Bush, "As We May Think," *Atlantic Monthly* 176 (July 1945): 101–18, probably the earliest published suggestion of the possibility of automated microfilm retrieval; Ralph R. Shaw, "The Rapid Selector," *Journal of Documentation* 5 (1949): 164–71; Alfred S. Tauber, "A Review of Microphotographic Techniques and Graphic Storage and Retrieval Systems," *Drexel Library Quarterly* 5 (1969): 234–40; Rodd S. Exelbert and Mitchell Badler, "Automatic Information Retrieval: A Report on the State of the Art," *Information and Records Management* 8 (May 1974): 23–29, 33–34; Stephen H. Wilcox, "Kodak MIRAcode II Products: Concepts and Applications," *Micrographics Science 1973*, ed. Chenevert, pp. 13–28; Arthur A. Teplitz, "Computer-Controlled Retrieval: A Primer," *Journal of Micrographics* 5 (1971): 35–40; Robert M. Landau, "New Economic Factors in the System Integration of Computer Terminal Online Retrieval Systems and Large Microform Data Banks," *Journal of Micrographics* 5 (1972): 125–30; Leon Bloom, "Information Retrieval Using Micrographics," *Journal of Micrographics*

8 (1974): 55–62; Carl F. J. Overhage and J. F. Reintjes, "Project Intrex: A General Review," *Information Storage and Retrieval* 10 (1974): 157–58; Robert J. Campbell, "Automated Microfilm Retrieval: A Refresher Course," *Modern Office Procedures* 25 (Oct. 1980): 63–64, 68; Robert M. Hayes, Online Microfiche Catalogs," *Journal of Micrographics* 13 (March/April 1980): 15–33, 58–63; "Computer-Assisted Retrieval Comes of Age," *Information and Records Management* 14 (May 1980): 42; Richard J. Skowitz, "Selection Criteria for Automated Retrieval Devices," *Journal of Micrographics* 13 (Sept./Oct. 1979): 41–43; Patricia M. O'Donnell, "Patentability Searching Using CCMSS," *Journal of Micrographics* 13 (Sept./Oct. 1979): 11–15.

Chapter 5

1. As noted in Chapter Two, copiers were considered a facet of word processing in the early 1970s. Facsimile technology, a form of electronic mail described in Chapter Six, is closely related to copier technology. The relationship between micrographics and reprographics dates from the early twentieth century. See, for example: R. C. Binkley, *Manual on Reproducing Research Materials* (Ann Arbor, 1936) and the classic W. R. Hawken, *Copying Methods Manual* (Chicago, 1966). Recent books dealing with the reproduction of research materials routinely consider micrographics a specialized variant of reprographics technology; examples include: Peter G. New, *Reprography for Librarians* (London, 1975) and C. G. LaHood, Jr. and R. C. Sullivan, *Reprographic Services in Libraries: Organization and Administration* (Chicago, 1975).
2. On the development of electrostatic technology, see: John H. Dessauer and Harold E. Clark, eds., *Xerography and Related Processes* (London, 1965); Dessauer, *My Years with Xerox: The Billions Nobody Wanted* (New York, 1971); R. M. Schaffert, *Electrophotography* (London, 1965); C. J. Young and H. G. Greig, "Electrofax: Direct Electrophotographic Printing on Paper," *RCA Review* 15 (1954): 469–84; W. T. Myers, "Recent Advances in Electrophotography," *TAPPI* 53 (1970): 442–46; William A. Sullivan, "Copier Design: A Systems Approach," *TAPPI* 57 (1974): 80–83.
3. Copiers and duplicators are regularly the subject of articles in trade and business publications. Examples include: "Convenience Copiers are Looking Better than Ever," *Modern Office Procedures* 25 (April 1980): 89–95; Ken Fukae, "What Computer Technology is Doing for Copying Machines," *Office* 91 (March 1980): 104; "New Features Improve Productivity in 1980," *Modern Office Procedures* 25 (Jan. 1980): 113–18; Jeffrey S. Prince, "The Case for In-House Reproduction," *Administrative Management* 41 (Sept. 1980): 45–46, 54; Paul J. Nickels, "Repro Update: Rebirth of a Stable Technology," *Modern Office Procedures* 24 (Oct. 1979): 37–42; C. C. Wilson, "Development of IBM Series III Copier—A Study in Change," *IAS Annual Meeting Proceedings* (New York, 1977): 291–4; A. Carusone, P. D. Clark and B. L. Tatom, "Automatic Diagnostics in a Processor-Controlled Copier/Printer," *IBM Technical Disclosure Bulletin* 21 (1979): 4397–98; Peter D. Foster, "Reduction of Wet Copy and Blooming in High Speed Electrofax Copiers," *TAPPI* 61 (1978): 105–7; M. J. Langdon, "The Xerox 6500 Color Copier: Design and Development," *IAS Annual Meeting Proceedings*, pp. 428–31.

4. W. A. Walshe, "New Copiers Mean More Flexibility for WP/AS," *Word Processing World* 6 (July 1979): 18–24; R. Clark Dubois, "Intelligent Copiers Will be Key Units in Future Offices," *Office* 89 (May 1979): 98, 106; Robert Conrad and Lawrence Dietz, "Hybrid Copiers are Learning More," *Data Management* 17 (May 1979): 54–55; David A. Rivers, "Intelligent Copiers Revolutionize Document Production," *Word Processing Systems* 7 (July 1980): 21–26; "OCR, Facsimile, Intelligent Copiers: Will They Merge?" *Word Processing Systems* 7 (April 1980): 28, 56; "Intelligent Copying is Growing Up," *Modern Office Procedures* 25 (Feb. 1980): 58–60; D. L. Buddington, "Copier Having Facsimile Scan/Print Function," *IBM Technical Disclosure Bulletin* 23 (June 1980): 75–78.
5. W. A. Lippold, "Phototypesetters for the Office: Their Selection and Use," *Administrative Management* 37 (Feb. 1976): 38–48; George A. Boucher, "Typesetting—A Must for Multiple-Page, Multiple-Copy Documents," *Office* 92 (Oct. 1980): 136–38; D. Martin, "Computer Typesetting and Information," *Journal of Documentation* 28 (1972): 247–49; Jeffrey S. Prince, "The Growing Sophistication of Photocomposers," *Administrative Management* 41 (Aug. 1980): 38–41, 81; "Photocomposers: Growing More Sophisticated All the Time," *Word Processing Systems* 7 (April 1980): 19–24; R. W. Bemer and A. R. Shriver, "Integrating Computer Text Processing with Photocomposition," *IEEE Transactions on Professional Communication* 16 (1973): 92–96; P. F. Mack, "Lower Composition Costs through Optical Scanning and Photocomposition," *IEEE Transactions on Professional Communication* 18 (1975): 279–82; P. J. Boehm, "Software and Hardware Considerations for a Technical Typesetting System," *IEEE Transactions on Professional Communication* 19 (1976): 15–19.
6. Linda Strub, "Combining Word Processing and Typesetting Enhances Productivity, Cuts Costs," *Word Processing Systems* 7 (June 1980): 67; T. S. Falletta, "Interface of Word Processing with Typesetting," *Scholarly Publishing* 14 (1980): 171–78; Elizabeth Ferrarini, "Phototypesetters and Word Processors—How to Solve the Incompatibility Problem," *Office* 91 (June 1980): 26, 30, 42, 56.
7. The definitive discussion is Henry B. Freedman et al., *An Exploratory Assessment of Computer-Assisted Makeup and Imaging Systems (CAMIS)* (Springfield, Va., 1980).

Chapter 6

1. The large number of recent publications about electronic mail and message systems reflects their popularity with office systems analysts. James Martin, *The Wired Society* (Englewood Cliffs, N.J., 1978) is certainly the most enthusiastic overview. Other useful surveys include: Edwin E. Mier and J. Peter Schmader, "Data Communications in the Office," *Data Communications* 9 (April 1980): 50–76; W. David Gardner, "A Look at Electronic Mail," *Output* 1 (April 1980): 38–41; R. J. Raggett, "A Happy Marriage Predicted for Computers, Phones," *Telephony,* Jan. 21, 1980, pp. 37–38; Frederick W. Miller, "Electronic Mail: A Smart Way to Do Business?" *Infosystems* 26 (Dec. 1979): 28–32; Martin Lasden, "Will You Love Electronic Mail or Hate It?" *Computer Decisions* 11 (Dec. 1979): 47–60; Howard Anderson, "Electronic Mail for the 80s," *Office* 90 (Nov. 1979): 18–22; J. N. Bruno, "Electronic Mail: It

Gets There Fast," *Administrative Management* 40 (Sept. 1979): 28–29; 64–70; W. A. Walshe, "Electronic Mail Diversifies with Technological Innovation," *Word Processing World* 6 (April 1979): 15–22; W. Flohrer, "Electronic Mail: Communication of Tomorrow," *Telephony,* March 12, 1979, pp. 89–96, 116–17; Victor Lederer, "Electronic Mail: The Ways to Go," *Administrative Management* 39 (Aug. 1978): 36–38; Robert J. Potter, "Electronic Mail," *Science* 195 (March 1977): 1160–64; E. B. Carne, "Telecommunications: Its Impact on Business," *Harvard Business Review* 50 (July/Aug. 1972): 125–30. On the concept of "speech mail," see W. A. Saxton and Morris Edwards, "Communication: Productivity Tool for the 1980s," *Infosystems* 26 (Aug. 1979): 70.

2. The significant national implications of electronic mail and related office information systems are discussed in Marvin Sirbu, "Automating Office Communications: The Policy Dilemmas," *Technology Review* 80 (Oct. 1978): 50–57. See also: Henry B. Freedman, "Laser Printing On-Demand: Evolution and Implications," *SPIE Journal* 169 (1979): 17–21; Hugh Collins, "Forecasting the Use of Innovative Telecommunications Services," *Futures* 12 (April 1980): 106–12; Jean Voge, "Information and Information Technologies in Growth and the Economic Crisis," *Technological Forecasting and Social Change* 14 (June 1979): 1–14. At the time this chapter was written, the U.S. Postal Service was in the process of seeking final approval for an electronic mail service to be called "ECOM." The proposed service has been strongly criticized by the private sector. Tom Alexander, "The Postal Service Would Like to be the Electronic Mailman, Too," *Fortune,* June 18, 1979, pp. 92–100 provides a good summary of the controversy. See also: Henry Geller and Stuart Brotman, "Electronic Alternatives to the U.S. Postal Service," *Direct Marketing* 41 (Jan. 1979): 67–75, 81; Geller and Brotman, "The Electronic Mailman," *Across the Board* 16 (July 1979): 35–43; W. F. Bolger, "Electronic Mail: Room for Partnership?" *Telephony,* July 16, 1979, pp. 28–30; "Post Office is Set for Computer Mail, but Will Seek Some Key Changes," *New York Times,* Feb. 23, 1980, p. 28; "Postal Service Board Decides on '82 Start for Electronic Mail," *Wall Street Journal,* Aug. 18, 1980, p. 3. In the meantime, the USPS is participating in Intelpost, a facsimile-based service for international communications. For a description, see Phil Hirsch, "USPS Begins Two-Way Intelpost with Canada," *Computerworld,* Oct. 6, 1980, p. 15; "Generations of Electronic Message Services," *Satellite Communications* 3 (June 1979): 24–28; "Who Will Fax the Mail?" *Datamation* 25 (Aug. 1979): 75, 77, 79.

3. "Aetna Tries Using TV to Cut Travel Costs," *Business Week,* March 6, 1978, pp. 34–36; "ARCO's $20 Million Talk Network," *Business Week,* July 7, 1980, pp. 81–82. On the relationship of telecommunications, travel and office location, see: Roger Pye, "Office Location: The Role of Communications and Technology," *Spatial Patterns of Office Growth and Location,* ed. P. W. Daniels (London, 1979), p. 239–55; R. Pye and E. Williams, "Teleconferencing: Is Video Valuable or is Audio Adequate?" *Telecommunications Policy* 1 (1977): 230–41; R. M. Haig, "Toward an Understanding of the Metropolis," *Quarterly Journal of Economics* 40 (1926): 402–34, an early study of the relationship of communications to office location.

4. For comparatively nontechnical discussions of fiber optics, see: G. Bylinsky, "Fiber Optics Finally Sees the Light of Day," *Fortune,* March 24, 1980, pp. 110–20; E. M. Ferrarini, "Fiber Optics: Telecommunications' Next Connec-

tion," *Administrative Management* 41 (March 1980): 53–56; A. A. Boraiko, "Fiber Optics—Harnessing Light by a Thread," *National Geographic* 156 (1979): 516.
5. On the technology and application of value-added, packet-switched networks, see S. L. Mathison, "Commercial, Legal, and International Aspects of Packet Communications," *Proceedings of the IEEE* 66 (1978): 1527–39; G. A. Duffy, "The Principles of Packet Switching," *Administrative Management* 39 (Aug. 1978): 32–33; B. D. Wessler and R. B. Hovey, "Public Packet-Switched Networks," *Datamation* 20 (July 1974): 85–87; K. Kummerle and H. Rudin, "Packet and Circuit Switching—Cost Performance Boundaries," *Computer Networks* 2 (1978): 3–17; E. G. Manning, "Datagram Service in Public Packet-Switched Networks," *Computer Networks* 2 (1978): 79–83; M. J. Fischer et al., "Large Scale Communication Networks—Design and Analysis," *Omega* 6 (1978): 331–40; B. Combs, "Tymnet—Distributed Network," *Datamation* 19 (July 1973): 40–43; M. P. Beere and N. C. Sullivan, "Tymnet—Serendipitous Evolution," *IEEE Transactions on Communications* 20 (1972): 511; Marshall Clinton, "Packet Switching Networks: Their Technology and Costs," *Online* 2 (July 1978): 51–54; Ray W. Sanders, "Comparing Networking Technologies: Ten Years Ago, Packet Switching Was Seen as the Most Reasonable Technology for Data Communications Networks," *Datamation* 24 (1978): 88–94.
6. For an excellent overview, see Pier L. Bargellini, "Commercial U.S. Satellites," *IEEE Spectrum* 16 (Oct. 1979): 30–37. Other useful sources include: "The Business Promise of Satellite Systems," *Industry Week*, Oct. 1, 1979, pp. 92–94; Ronald A. Frank, "How SBS Service Stacks Up," *Datamation* 26 (Oct. 1980): 54–56; Nicolas Mokhoff, "Communications and Microwave," *IEEE Spectrum* 17 (Jan. 1980): 38–43. James Martin, *Communications Satellite Systems* (Englewood Cliffs, N.J., 1978) describes the technology of satellite communication. See also: I. M. Jacobs, R. Binder and E. V. Hoversten, "General Purpose Packet Satellite Networks," *Proceedings of the IEEE* 66 (1978): 1448–67; R. W. Williams, "Experimenting with Satellite-Linked Computer Networks," *Hewlett-Packard Journal* 29 (1978): 27–32; B. I. Nelson, "Satellite Communications for Information Networking," *Proceedings of the American Society for Information Science* 9 (1972): 175–87; G. R. Thoma, "Performance Evaluation of a Satellite Linked Experimental Network," *IEEE Transactions on Aerospace and Electronic Systems* 16 (1980): 771–81.
7. R. M. Metcalfe and D. R. Boggs, "Ethernet—Distributed Packet Switching for Local Computer-Networks," *Communications of the Association for Computing Machinery* 19 (1976): 395–404; D. R. Boggs and R. M. Metcalfe, "Ethernet Packet Transport for Distributed Computing," *Proceedings of the Spring Compcon 78* (New York, 1978), p. 152; "The Office of the Future Gets a Common Voice," *Business Week*, May 26, 1980, pp. 57–58; "Xerox's Bid to Unlock the Office of the Future," *Business Week*, Dec. 24, 1979, p. 47; Ivan T. Frisch, "Planning Local-Area Nets Harder than It Looks," *Computerworld*, Nov. 24, 1980, pp. 3, 4, 6; Brad Schultz, "IEEE Group Drafting Local Net Standard," *Computerworld*, Aug. 25, 1980, pp. 49–50; W. A. Saxton and Morris Edwards, "Networks for the Office," *Infosystems* 26 (June 1979): 108–10; "NBS Experimenting with Ethernet Packet Switching," *Data Communications* 6 (Feb. 1977): 20.
8. The best survey of the facsimile and related technologies is Daniel M. Costigan,

Electronic Delivery of Documents and Graphics (New York, 1978). See also William Saffady, "Facsimile Transmission for Libraries: Technology and Applications Design," *Library Technology Reports* 14 (1978): 445–532, which includes a bibliography of the literature through 1977.
9. In addition to the sources cited in the preceding footnote, see A. N. Netravali and F. W. Mounts, "Ordering Techniques for Facsimile Coding: A Review," *Proceedings of the IEEE* 68 (1980): 796–807.
10. R. Hunter and A. H. Robinson, "International Digital Facsimile Coding Standards," *Proceedings of the IEEE* 68 (1980): 854–67.
11. See, for example: A. M. Noll and J. P. Woods, "Use of Picturephone in a Hospital," *Telecommunications Policy* 3 (1979): 29–36. For an interesting study of the psychological and sociological implications of telecommunications technology, see: Edward M. Dickson and Raymond Bowers, "Human Response to Video Telephones," *A Technology Assessment Primer,* eds. Leon Kirchmayer, Harold Linstone, and William Morsch (New York, 1975), pp. 148–60.
12. The amount of literature on videoconferencing is small but growing. Recent articles include: J. P. Duncanso and A. D. Williams, "Video Conferencing—Reactions of Users," *Human Factors* 15 (1973): 471–85; E. F. Brown et al., "Some Objective and Subjective Differences between Communication over Video-Conferencing Systems," *IEEE Transactions on Communications* 28 (1980): 759–64; T. Watanabe, "Teleconferencing in Japan—Use of Audio Conference Systems and Evolution Towards Video," *Telecommunications Policy* 3 (1979): 290–96; M. Kikuchi and M. Yoshikawa, "New Video Conference System," *Japan Telecommunications Review* 22 (1980): 112–19; N. Mokhoff, "They Said It Couldn't Be Done: The Global Video Conference," *IEEE Spectrum* 17 (Sept. 1980): 44.
13. On the technology and applications of video disks, see: Maria Savage, "Beyond Film: A Look at the Information Storage Potential of Videodiscs," *Bulletin of the American Society for Information Science* 7 (Oct. 1980): 26–30; "Videodisc Information Capacity," *American Libraries* 11 (March 1980): 174; "The Videodisc Revolution: a New Medium of Information," *Futurist* 11 (Oct. 1977): 311–12; D. S. McCoy, "RCA Selectavision Videodisc System," *RCA Review* 39 (March 1978): 7–13; A. Firester et al., "Optical Readout of the RCA Videodisc," *RCA Review* 39 (Sept. 1978): 392–426; Peter W. Bogels, "System Coding Parameters, Mechanics, and Electromechanics of the Reflective Videodisc Player," *Journal of the SMPTE* 86 (1977): 144–45; George C. Kenney, "Special Purpose Applications of the Optical Videodisc System," *IEEE Transactions on Consumer Electronics* 22 (1976): 327–36; Kent D. Broadbent, "Review of the MCA Disco-Vision System," *Information Display* 12 (1976): 12–19; "Videodiscs, a Three-Way Race for a Billion Dollar Jackpot," *Business Week,* July 7, 1980, pp. 72–80.
14. The question is examined by Gerard O. Walter, "Will Optical Disk Memory Supplant Microfilm?" *Journal of Micrographics* 13 (July/Aug. 1980): 29–34; S. Suthasinekul, "Microfilm vs. Optical Disk as a Storage Medium for Document-Retrieval and Dissemination," *Journal of the American Society for Information Science* 17 (1980): 100–102.
15. See Gerald Tomanek, "Implementation of Electronic Mail in a Telephone System," *AFIPS Proceedings* 49 (1980): 527–29.

16. For a summary of the most commonly encountered features, see Walter E. Ulrich, "Implementation Considerations in Electronic Mail," *AFIPS Proceedings* 49 (1980): 489–91; Harold E. O'Kelley, "Electronic Message System as a Function in the Integrated Electronic Office," *AFIPS Proceedings* 49 (1980): 499–505.
17. The results of one survey are reported in Jeffrey B. Holden, "Experiences of an Electronic Mail Vendor," *AFIPS Proceedings* 49 (1980): 493.
18. See Robert B. White, "A Prototype for the Automated Office," *Datamation* 23 (April 1977): 83–90.
19. The most comprehensive discussion of teleconferencing is Murray Turoff and Roxanne S. Hiltz, *The Network Nation: Human Communication via Computer* (Reading, Mass., 1978), which includes a bibliography.

Chapter 7

1. For a description of a model integrated office, see Charles I. Norris, "Integrated Office Systems: The Paperless Office," *Journal of Micrographics* 13 (May/June 1980): 23–26; H. T. Smith, "Project '80: Integrating Information Systems," *Management World* 9 (Jan. 1980): 9, 32ff; A. Mokler, "About Implementation of Knowledge-Based and Integrated Office Information Systems," *Angewandte Informatik* 12 (1979): 542–49.
2. R. Jurk, "Work Desk and Communication Facilities in the Office of the Future," in *Electronic Text Communication,* ed. W. Kaiser (Berlin, 1978), pp. 458–67; W. Postl and W. Woborschil, "Integrated Office Communication for Text and Graphics," *NTG-Fachber* 67 (1979): 254–62.
3. See John J. Brennan, "Business Communications," a special advertising supplement prepared by Quantum Science Corporation in *Forbes,* June 23, 1980.
4. Melvin J. Kirschner, "Word Processing—Friend or Foe of Micrographics?" *IMC Journal* 16 (Third Quarter, 1980): 8–10; Marguerite Zientara, "WP-COM Combination Seen Dynamic Duo," *Computerworld,* May 19, 1980, p. 102; Lincoln Hallen, "Integrating Micrographics into Future Office Systems," *Journal of Micrographics* 13 (March/April 1980): 71–77; "Word Processing and Micrographics: Natural Allies for Storage and Retrieval System," *Word Processing Systems* 6 (Oct. 1979): 22–28; Alan S. Linden, "Word Processing and Micrographics," *Journal of Micrographics* 11 (Nov./Dec. 1977): 75–76.
5. Deborah S. Aframe, "The Congressional Research Service: A Case Study," *Journal of Micrographics* 12 (July/Aug. 1979): 371–75; A. S. Linden, "Congressional Research Retrieval Combines Movable WP Microfiche from Output," *Data Management* 17 (Nov. 1979): 14.
6. Studies of the impact of electronic technology on office workers are just now emerging. See, for example: F. Delorme, "Office Automation—Some Consequences for the Office-Work-Force," *Relations Industrielles/Industrial Relations* 29 (1974): 513–40; Dean J. Champion, "Some Impacts of Office Automation Upon Status, Role Change and Depersonalization," *Sociological Quarterly* 8 (1967): 71–84; M. Magnus, "Office Automation, Personnel and the New Technology," *Personnel Journal* 59 (Oct. 1980): 815–23. For a general discussion of the impact

of automation on worker attitudes, see: John T. Dunlop, *Automation and Technological Change* (Englewood Cliffs, N.J., 1962), especially pp. 43–65; and Louis E. Davis and Albert E. Cherns, eds., *The Quality of Working Life,* vol. 1 (New York, 1975), especially pp. 220–257.

Index

acoustic coupler, 59, 76, 121
A.B. Dick System 200 Record Processor, 98
adding machines, 3
administrative secretaries, 45, 46
ADSTAR systems, 116
Aetna, 159
agricultural sector, 9, 10
Alden Electronic and Impulse Recording Company, 172
American Autotypist, 31
American Satellite Company, 159
American Standard Code for Information Interchange (ASCII), 57, 58, 62, 86, 121, 179, 184–5
American Telephone and Telegraph Company, 155, 176
AMI, 79
AM International, 159
amplitude modulation, 164, 171
AM Varityper Composer, 141
Antiope system, 71
aperture cards, 18, 89, 91, 96, 116, 118, 124
APL, 57, 62, 84
Apple computers, 81
ARPANET, 186
ASCII. *See* American Standard Code for Information Interchange.

Associated Press, 69
asynchronous transmission, 56, 58, 86
automated text editing, 16, 17, 19, 20, 21, 31–50, 53, 55, 77, 85, 113, 141, 145
Automatic Send-Receive (ARS) terminal, 55
"back-office" operations, 3, 5, 52, 66
BASIC, 84, 196
Basic Four, 196
batch processing, 52, 64–5, 89, 112–113
bar charts, 73
bauds, 58
Baudot code, 58, 181
Bell and Howell Microx System II, 98
Bell System, 155–6
Bell Telephone Laboratories, 176
Bibliographic Retrieval Services, 69
binary coded decimal (BCD), 58
bits, 56, 58, 79
blip encoding, 94, 116, 117
blue collar workers, 8, 10–11
Bolt, Berenak, and Newman, 157
Boolean logic, 126
British Broadcasting Company, 71
byte, 79
calculators, 6, 13, 78
calendar management, 198

235

CAMIS technology, 149–51
Canon, 140
capital investment, 13
CAR. *See* computer-assisted retrieval.
cassettes, 23, 25, 28, 31, 40–1, 42, 44, 145, 147
cathode ray tube (CRT) displays, 29, 34–7, 43, 49, 54–64, 89, 104–106, 121, 133, 143–4, 145, 146–7, 148, 150, 185, 195
CCITT no. 2 code, 181
Ceefax teletext service, 71
Central Intelligence Agency, 116
centralized dictation systems, 28–9, 30–1
character-oriented message transmission, 179–91
Citibank, 190
Citicorp, 186
clerical workers, 5–6, 14, 18, 151
coaxial cable, 20, 158, 160–1
COBOL, 66, 84
COM. *See* computer-output microfilm.
COMET electronic message system, 157
Comité Consultatif International Télégraphique et Téléphonique (CCITT), 171
communicating word processors, 183–5, 201
COMp 80/2, 110, 143, 201
computer-assisted retrieval (CAR), 4, 18, 91, 95, 114–26, 172, 174, 194, 200, 204
computer-based message systems (CBMS), 20, 154, 185–90
Computer Corporation of America, 157
computer graphics, 17, 71–5, 107
computer-input microfilm (CIM), 114
computer-output microfilm (COM), 18, 52, 89, 91, 104–14, 116, 117, 138, 172, 200–201
computer terminals, 7, 17, 52–64, 77, 89, 123, 125
computers, 3, 5, 7, 15, 16–17, 18, 19, 39, 51–87, 91, 104–26, 141, 149–51, 160, 185–90, 199–201

Comsat General, 159
Continental Telephone, 159
copiers, 4, 6, 13, 15, 19, 103, 128–41
Correspondence, 58
CP/M, 84
Data Access Arrangement (DAA), 59
data banks, 67–71
data base management systems, 7, 64–7, 72, 123
data bases, 7, 17, 18, 64–71, 113–14, 116, 157
Dataphone Digital Service, 155
Datapoint, 196
data processing, 40, 49–50, 56, 72, 75, 196, 197, 201
DEC PDP-1, 76
desk-top dictation systems, 23–5, 28, 30
Diablo Hy-Type printer, 33, 35, 81
DIALOG, 69–71
diazo duplication, 101–2, 106, 137
dictation systems, 3, 15, 20–30
Didon teletext service, 71
Digital Equipment Corporation, 76, 86, 161, 194
Digital PABX, 186, 198
Digital Research Incorporated, 84
direct-entry typesetters, 145–6, 150
direct-impression typesetting, 141
disk storage, 41–2, 64, 76, 112, 113, 114, 199
diskettes. *See* floppy disks.
display-oriented text-editing systems, 34–7, 40, 43, 85, 145, 200
display terminals. *See* cathode ray tube displays.
distributed data processing, 40, 76
document assembly, 45
document conveyors, 3, 6
documentation for automated office systems, 202–5
dot matrix printers, 61–2, 81
Dow Jones News/Retrieval Service, 69
dry silver microfilm, 97–8, 106
"dumb" terminals, 53
duplicators, 19, 128–41
Eastman Kodak Company, 116

electrofax process, 128–9, 133, 165
electrolytic recording, 165
electronic blackboard, 176
electronic mail and message systems (EMMS), 7, 15, 17, 19–20, 40, 153–91, 201
Electronics Industry Association, 59, 64
Electro-Optical Mechanisms, 101
electropercussive recording, 165
electrosensitive recording, 165
electrostatic process, 128–29
Energy Conversion Devices, Inc., 100
EOM 6100 microfilm camera, 101
Ethernet, 161, 194
Extended Binary Coded Decimal Interchange Code (EBCDIC), 58, 184, 185
Exxon Corporation, 161
facsimile, 20, 90, 154, 157, 159, 162–72, 185, 194, 196
Fairchild Industries, 159
fiber optics, 19, 130, 138, 140, 156, 162
firmware, 38, 40, 56, 80
flat panel displays, 54, 195
floppy disks, 38, 39, 40–2, 44, 81–3, 122, 145, 147, 200
form slide, 106, 111, 117
FORTRAN, 66, 84
FOSDIC, 114
FR-80 COM recorder, 114
French Telegraph Cable Company (FTCC), 180
frequency modulation, 164, 171
Friden Flexowriter, 31
"front office" operations, 5, 51, 66
full-duplex transmission, 58
global changes, 44
Grafix I, 114
Graphic Scanning Corporation, 172
Graphic Sciences, 164
graphics display terminals, 74
graphics plotter, 74, 81
Graphnet, 157, 172, 182
Grosch, Herbert R. J., 75–6
Grosch's law, 75–6
Gross National Product (GNP), 8

half-duplex transmission, 58
HERMES electronic message system, 157
histograms, 73
Holiday Inns, 176
IBM, 86, 138, 159
IBM Correcting Selectric, 38, 43
IBM Magnetic Card Selectric Typewriter, 32–3, 36, 38, 40–1, 139
IBM Magnetic Tape Selectric Typewriter (MT/ST), 3–4, 32, 40
IBM Office system/6, 35, 139
IBM Selectric Composer, 141
IBM Selectric Typewriter, 3, 32–5, 46, 47, 55, 58, 61
IBM 3800, 110, 129
IBM 6670 Information Distributor, 139–40
IBM System 6:5, 30
image digitizers, 81
image-oriented message transmission, 162–79
impact printers, 60–2, 81
Info Master Service, 181–2
Information International, Inc., 110, 114, 143
ink-jet printer, 35
Institution of Electrical and Electronic Engineers (IEEE), 161
integrated information systems, 193–202
integration of technologies, 6–7
Intel, 79, 80, 85, 161, 194
intelligent copiers, 5, 15, 19, 35, 138–41, 150, 160, 194, 201
"intelligent" microfilm equipment, 91–5
intelligent microfilmers, 194
Intelligent Query System, 174
intelligent copiers, 177
intelligent terminals, 17, 53, 85
intelligent typewriters, 16, 46–7
interfacing automated office systems, 198–201
International Telephone and Telegraph, 172
ITT Faxpak, 157, 172
ITT World Communications, 180

Japan, 9, 10, 129
Keyboard Send-Receive (KSR) terminals, 54-5, 64
keypunch, 52
key word indexes, 123-4
Konishiroku, 140
labor productivity, 8-9, 13
large-scale integration (LSI), 17, 76
laser recording, 106, 110-112, 129, 138, 144, 165, 166
laser scanning, 162
LegisLate, 71
light-emitting diode (LED) display, 37, 117
line printers, 104, 107-110, 111, 112, 199
list management, 45
Lockheed Retrieval Service, 69-71
longhand writing, 22
Magazine Index, 70
Magnavox video disk system, 178
magnetic cards, 40-1, 42, 44, 139
magnetic tape, 52, 65, 76, 114, 200
Mailgram, 20, 154, 182-3, 185, 186
management information systems, 64-7
managers, 1, 3-6, 12, 14-15, 20, 46, 64, 67, 154, 205-6
　work patterns, 5
manufacturing sector, 9-11
mass storage systems, 65
Massachusetts Institute of Technology, 116
Matsushita, 178
MCI Communications, 158
MCA videodisk system, 178
Mead Data Central, 69
MEDLINE, 69
memory typewriters, 16, 46-7
microcassettes, 25, 31
microcomputers, 17, 38, 55, 76-87, 120, 194
microfacsimile, 172-5, 201
microfiche, 16, 18, 89, 92, 96, 97-100, 101, 104, 106-7, 108, 110, 114, 116, 117, 118, 126, 138, 172-3
microfilm, 16, 89, 96, 104, 106-7, 108, 114, 117, 124-5, 150, 172

microfilm cameras, 3, 18, 92, 96, 116, 132
microfilm card jackets, 101, 102
microfilm cartridges and cassettes, 89, 91, 117, 124, 126
microfilm duplicators, 100, 101-2
microfilm jackets, 18, 89, 91, 98, 101, 102, 118
microfilm retrieval units, 92, 94-5, 100
microfilm viewing equipment, 3, 101, 102-3
microforms, 18, 20, 89, 91, 111, 172
micrographics, 3, 4, 15, 17-18, 89-126, 127, 137, 167, 172, 201
microimages, 89
MicrOvonics, File, 98-100
microprocessors, 17, 18, 19, 24, 39, 40, 50, 53, 61, 62, 77, 79, 85, 91, 92, 94, 95, 100, 117, 122, 132, 137, 145, 150, 161, 170, 194, 195
microwave communications, 158, 159
Mills, C. Wright, 3
mimeography, 136-7
minicassettes, 25, 31
minicomputers, 4, 17, 39, 50, 75, 104, 120, 122
mini-floppy disks, 41, 82
Minolta, 140
Miracode®, 116
3M Company, 200
3M 1050 microfiche camera, 97-8
modems, 59, 86, 121, 155
Mostek, 79
motorized files, 90
Motorola, 79
multi-function systems, 7, 85, 196-8
National Library of Medicine, 69
National Micrographics Association (NMA), 110
NEC Spinwriter, 33, 35, 81
New York Times Information Bank, 69
NEXIS, 69
Nite Cast, 182
non-impact printers, 60-2, 81
OCR. See optical character recognition.

OCR-B type font, 114
OCR reader, 47–8, 147, 149
"office of the future," 2
offset duplication, 136
one-line text-editing displays, 37–8, 43
online computing, 52–71, 72, 89, 112–114, 120, 123, 153
online information systems, 17
On-Tyme electronic mail system, 172
open-plan office, 4
operating systems, 83–4
optical character recognition (OCR), 47–8, 114, 154, 179, 183–5, 201
optical disks, 65, 176–9
ORBIT, 69
packet switching, 157
page printers, 110–112, 129, 199
paper files, 90, 98
paperless information systems, 5, 89
Pascal, 84
personal computers, 78
phase modulation, 164
Philips videodisk system, 178
photocomposition, 48–9, 148–9, 150
photoplastic microfilm, 98
phototypesetting, 141–51
picture elements, 162
Picturephone, 20, 175
Pioneer Electronics videodisk system, 178
planetary cameras, 92–3, 100, 101
Planning Research Corporation, 172–4
plasma displays, 37, 54
PL/1, 66
portable dictation recorders, 25–6, 28
"post-industrial society," 12
Prestel system, 71
Prime Computer, 196
primary sector, 9–10
productivity, 2, 5–6, 8–13, 29, 134–6, 190
programmable calculators, 78
Project Intrex, 116
Public Affairs Information Service, 70
punched cards, 52
query language, 67, 68
QUME Corporation, 33, 35, 81

Radio Shack TRS-80, 81, 86
Ragen MRS-95, 117, 175
random-access memory (RAM), 63, 80, 81, 92
random-access microfiche systems, 118
RCA Global Communications, 175, 180
RCA videodisk system, 178
raster scanning, 175
read-only memory (ROM), 38, 80, 92, 122
read/write heads, 42
readers and reader-printers, 101, 102–3, 116, 117, 126
real-time information systems, 112–114, 123
Receive-Only (RO) terminals, 55, 64, 81
recorder, in dictation systems, 23, 24, 28
Redactron, 41
RediList Service, 183, 186
reprographics, 3, 15, 19, 127–51, 201
robotics, 11
ROLM Telecommunications, 186
rotary cameras, 92–3, 101
RS-232C, 59, 64, 86
RS-499, 59
run-length encoding, 168–70
Satellite Business Systems, 159
satellite communications, 158–60, 194
S-100 bus, 80–1, 85
scrolling, 35, 110
secondary sector, 9–11
services sector, 9–12
shared logic text-editing systems, 39–40
shared processor text-editing systems, 38–40, 50
shared resource text-editing systems, 40
Siemens' Munich Research Center, 195
silver halide process, 95–6, 98, 106, 166
small business computers, 75–87, 121, 177
small-office microfilm (SOM), 18, 91, 100–103

"smart" terminals, 53, 63, 121
socio-behavioral factors in office automation, 205–6
software, 17, 20, 38, 40, 50, 56, 64, 65–7, 74, 80, 83–5, 116, 122–3, 124, 125, 189, 196, 200
Sony Corporation, 198
sorters, 3, 19, 135
source document microfilming, 18, 90–103, 110
Southern Pacific Communications, 172
SPC Speedfax, 157, 172
speech mail, 198
spirit duplication, 136–7
STAIRS, 69
stand-alone text-editing systems, 38–40, 85, 196
stenographer, 22–3
step-and-repeat cameras, 92–4
STORTEX Service, 182
synchronous transmission, 56
Synertek, 79
System Development Corporation, 69
tabulating machines, 3
"tank-type" dictation systems, 28–9
technological change, 12–13
telecommunications, 17, 86, 139, 147, 149, 182
teleconferencing, 190–1
Telefiche System, 172–4
telegraphy, 20
Telenet, 157, 186
telephone network, 20, 58, 155–6, 158, 159
teleprinters, 54–64, 89, 121
teletext services, 71
teletype, 3, 57
telex, 20, 58, 154–5, 157, 172, 179–83, 184–5, 186, 189, 196, 197, 198, 201
Teletype Model 32, 181
Teletype Model 33, 179
teletypesetter (TTS) tape, 146–7, 148
Telidon, 71
TERA, 175
Terminal Data Corporation, 174–5
terminals. See computer terminals.

tertiary sector, 9–12
Texas Instruments, 79
thermal silver paper, 166
thermally processed silver microfilm, 106, 137
time sharing, 39, 52–3, 74
total productivity, 8, 13
transceiver, 162
transcriber, in dictation systems, 23, 24, 26, 29
transparent electrophotography, 98
TRT Telecommunications, 180
turnkey systems, 85, 116, 120, 121, 122–3
TWX, 20, 154–5, 157, 172, 179–83, 184–5, 186, 189, 196, 197, 198, 201
Tyme-Gram, 183
Tymnet, 157, 172, 183, 186
typesetters, 19, 48–9, 141–51, 153
typewriter-based text-editing systems, 32–4, 36, 40, 43
typewriters, 3, 6, 13, 15, 31, 46, 141, 153, 184
typists, 23
United Press International, 69
updatable microfiche systems, 18, 98–100
U.S. Bureau of Labor Statistics, 7–8
U.S. Census Bureau, 74, 114
U.S. Department of Defense, 186
U.S. Federal Communications Commission, 159
U.S. Postal Service, 166
value-added networks, 157, 159, 172, 182
variable velocity scanning, 169
vesicular duplication, 101–102, 106, 137
video composers, 148–9
video conferencing, 20, 159, 175–9, 190
video disks, 65, 176–9
video technology, 20, 175–6
videotext services, 71
Videovoice system, 175
viewdata services, 71
voice communications, 153, 155

voice-operated relay (VOR) technology, 24
voice synthesizers, 81
Wang Intelligent Image Printer, 138
Wang Office Information System, 139
Walnut document storage and retrieval system, 116
WESTAR satellites, 159
Western Union, 159, 179–83
Western Union International, 180
white collar workers, 1, 6, 8, 9, 12, 13, 14–15, 64, 68
Winchester-type disks, 42, 83, 122
word, 79
word originator, 21–3, 26, 47
word processing, 5, 7, 15–16, 17, 19, 21–50, 56, 81, 85, 110, 141, 145–8, 149, 154, 160, 177, 183–5, 194, 196, 197, 199–201

word processing operators, 45
word processing service bureaus, 39
workforce composition, 9–12
xerography, 3, 4, 35, 110, 128, 129, 133, 150, 159, 165, 196
Xerox Corporation, 41, 86, 128, 161, 180, 194
Xerox 800 Series Electronic Typing System, 33
Xerox 850 text-editing system, 138
Xerox 5700, 140–1
Xerox 6500 color copier, 133
Xerox 9400, 110, 136
Xerox 9700, 110–11, 124, 138, 140, 200
Xerox XTEN satellite network, 159
Zilog, 79, 80, 161